Standard of Ministry of Water Resources of
the People's Republic of China

SL 326—2005
Replace DL 5010—92

Code for Engineering Geophysical Exploration of Water Resources and Hydropower

Drafted by:
Changjiang Institute of Survey, Planning, Design and Research of Changjiang Water Resources Commission

Translated by:
Changjiang Institute of Survey, Planning, Design and Research of Changjiang Water Resources Commission

China Water & Power Press
Beijing 2015

图书在版编目（CIP）数据

水利水电工程物探规程 = Code for Engineering Geophysical Exploration of Water Resources and HydropowerSL326-2005 Replace DL 5010-92：英文 / 中华人民共和国水利部发布. -- 北京：中国水利水电出版社，2015.1
 ISBN 978-7-5170-2958-8

Ⅰ. ①水… Ⅱ. ①中… Ⅲ. ①水利水电工程－地球物理勘探－技术操作规程－中国－英文 Ⅳ. ①TV698.1-65

中国版本图书馆CIP数据核字(2015)第035221号

书　名	Code for Engineering Geophysical Exploration of Water Resources and Hydropower SL 326—2005 Replace DL 5010—92
作　者	中华人民共和国水利部　发布
出版发行	中国水利水电出版社 （北京市海淀区玉渊潭南路1号D座　100038） 网址：www.waterpub.com.cn E-mail：sales@waterpub.com.cn 电话：(010) 68367658（发行部）
经　售	北京科水图书销售中心（零售） 电话：(010) 88383994、63202643、68545874 全国各地新华书店和相关出版物销售网点
排　版	中国水利水电出版社微机排版中心
印　刷	北京瑞斯通印务发展有限公司
规　格	140mm×203mm　32开本　6.875印张　240千字
版　次	2015年1月第1版　2015年1月第1次印刷
定　价	**408.00元**

凡购买我社图书，如有缺页、倒页、脱页的，本社发行部负责调换

版权所有·侵权必究

Introduction to English Version

Department of International Cooperation, Science and Technology of Ministry of Water Resources, P. R. China has the mandate of managing the formulation and revision of water technology standards in China.

Translation of the code from Chinese into English was organized by Department of International Cooperation, Science and Technology of Ministry of Water Resources, P. R. China in accordance with due procedures and regulations applicable in the country.

This English version of code is identical to its Chinese original SL 326—2005 *Code for Engineering Geophysical Exploration of Water Resources and Hydropower*, which was formulated and revised under the auspices of Department of International Cooperation, Science and Technology of Ministry of Water Resources, P. R. China.

Translation of this code is undertaken by Changjiang Institute of Survey, Planning, Design and Research of Changjiang Water Resources Commission.

Translation task force includes Zhang Jianqing, Lu Ernan, Zhang Zhi, Liu Runze, Cai Jiaxing, Kuang Bibo, Chen Shangfa, Li Shengqing, Ma Shengmin, Zhang Jinfeng, Chang Yong, Lu Jie, and Zheng Wei.

This code is reviewed by Li Guangcheng, Wang Yicheng,

Yang Dingyuan, Jin Hai, Meng Zhimin, Gu Liya, Cheng Xialei, Hao Jutao, and Tian Ziqin.

Department of International Cooperation, Science and Technology Ministry of Water Resources, P. R. China

Foreword

Modification, improvement and supplement are made for this code in accordance with the *Notification on Project Plan and Chief Development Organization on Preparation and Revision of Specifications Investigation and Design Technology of Water Resources and Hydropower Engineering in 2001* in WRHPDA Doc. No. 1 〔2001〕, the information on the last decade of execution and the status quo of technical development of geophysical exploration for DL 5010—92 *Code for Engineering Geophysical Exploration of Water Resources and Hydropower*, and the SL 1—2002 *Regulations on Preparing Technical Specifications for Water Resources*.

SL 326—2005 *Code for Engineering Geophysical Exploration of Water Resources and Hydropower* consists of 5 chapters, 35 sections and 3 annexes. This code mainly includes:

—Definition of terminologies, symbols and codes.

—Clear identification of methods and categories used in geophysical methods and techniques, application conditions and instrument specifications for each method, and determination of technical requirements for field experiments, observation, arrangement of survey grids, parameter measurements, records and assessment, data processing and interpretation, reports and plots, etc.

—Determination of integrated applications of geophysical methods in 22 areas, clear identification of methods and techniques employed in each application area, and proposal of specific requirements for data interpretation and accuracy.

—For this revision of DL 5010—92 *Code for Engineering Geophysical Exploration of Water Resources and Hydropower*, the deleted and added contents are detailed as follows:

—The deleted contents are task of geophysical exploration in Chapter 2, very-low-frequency in Section 3.1, microgravity survey in Section 3.3, explanatory provisions for related methods and techniques in Chapter 3, detection of pile foundation in Section 4.11, use and maintenance of instruments and equipment in Annex A, provisions for related operating procedures in Annex B, plot representation and legend in Annex F, safety regulations for seismic exploration blasting practices in Annex G, regulation on the safety and protection of radioisotopes and radiation devices in Annex H.

—The added contents are normative reference, terms, symbols and codes, the items in the chapter of geophysical method and technique including general provisions, resistivity imaging, transient electromagnetic method, controlled source-audio maganetotellurics, ground penetrating radar, Rayleigh wave, computerized tomography, etc. as well as the items in the chapter of integrated applications of geophysical methods including detection of hidden defects in dams and embankments, early prediction of tunnel construction, testing of grouting effect, quality detection of foundation bedrocks, concrete, concrete lining in cavities, bolt anchorage and cutoff wall, detection of rockfill (soilfill) density and bearing pressure on foundation, detection of the contact behavior between steel lining and concrete, quality detection of rockfill dam concrete face slab, etc.

This document replaces:

—*Code for Geophysical Exploration of Hydrological and Engineering Geology* (1982)

—DL 5010—92 *Code for Engineering Geophysical Exploration of*

Water Resources and Hydropower

This code is approved by the Ministry of Water Resources of the People's Republic of China.

This code is interpreted by Water Resources and Hydropower Planning and Design Administration of the Ministry of Water Resources.

This code is chiefly drafted by Changjiang Institute of Survey, Planning, Design and Research of Changjiang Water Resources Commission.

This code is jointly drafted by Yellow River Institute of Hydraulic Research of Yellow River Conservancy Commission, Yellow River Engineering Consulting Co., Ltd., Changjiang Survey Technological Research Institute of the Ministry of Water Resources, China Water Northeast Investigation, Design & Research Co., Ltd., China Water Beifang Investigation, Design & Research Co., Ltd., HydroChina Guiyang Engineering Corporation, HydroChina Chengdu Engineering Corporation, HydroChina Beijing Engineering Corporation.

This code is published and distributed by China Water & Power Press.

Chief drafters of this code are Xiao Boxun, Cai Jiaxing, Wang Bo, Sha Chun, Leng Yuanbao, Xiong Yonghong, Wei Yanjun, Zhang Zhi, Yu Caisheng, Zhang Jianqing, Yuan Jinghua, Guo Yusong, Wu Guangrong, Liu Kanghe, Wang Shunli, Cai Zhixuan, Song Zhengzong, Wu Dauan, Qian Shilong, and Zhang Xiyuan.

The technical responsible person of this code review meeting is Li Guangcheng.

The format examiner of this code is Dou Yisong.

Contents

Introduction to English Version
Foreword
1 General Provisions 1
2 Terms and Notations 4
   2.1 Terms 4
   2.2 Symbols 7
3 Geophysical Methods and Techniques 11
   3.1 General 11
   3.2 Electrical Survey 16
   3.3 Ground Penetrating Radar 50
   3.4 Seismic Survey 57
   3.5 Measurement of Elastic Waves 83
   3.6 Computerized Tomography 92
   3.7 Sonic Echo Exploration 99
   3.8 Radioactivity Survey 102
   3.9 Comprehensive Logging 110
4 Comprehensive Applications of Geophysical
   Methods 119
   4.1 Investigation of Overburden 119
   4.2 Investigation of Buried Geological Structures and
       Fractured Zones 121
   4.3 Investigation of Karst 125
   4.4 Investigation of Thickness of Weathered Rock Mass and Depth
       of Unloading Zones 127
   4.5 Investigation of Weak Interlayers 130
   4.6 Investigation of Landslide 133

4.7	Detection of Hidden Defects in Dams and Embankments	135
4.8	Advanced Prediction in Tunnel Excavation	138
4.9	Investigation of Groundwater	141
4.10	Detection of Environmental Radioactivity	144
4.11	Quality Detection of Foundation Bedrocks	147
4.12	Detection of Grouting Effect	151
4.13	Quality Detection of Concrete	154
4.14	Quality Detection of Concrete Lining in Cavern	157
4.15	Detection of Relaxation Zone Around Cavern	159
4.16	Quality Detection of Bolt Anchorage	160
4.17	Quality Detection of Cutoff Wall	163
4.18	Detection of Rockfill (Soilfill) Density and Bearing Capacity on Foundation	164
4.19	Detection of Contact Status Between Steel Lining and Concrete	166
4.20	Quality Detection of Rockfill Dam Concrete Face Slab	168
4.21	Measurement of Hydrogeological Parameters	169
4.22	Measurement of Rock and Soil Physical and Mechanics Parameters	171
5	Results and Reports for Geophysical Exploration	175
5.1	Compilation of Results and Reports	175
5.2	Review of Results	178
Annex A	Summary of Geophysical Exploration Applications	180
Annex B	Table of Physical Property Parameters	186
Annex C	Fundamental Equations and Calculation Charts	189

1 General Provisions

1.0.1 This code is formulated for regulating the technical requirements of engineering geophysical exploration, ensuring the quality of geophysical exploration results and giving full play to geophysical exploration in engineering survey of water resources and hydropower.

1.0.2 This code is applicable to engineering geophysical exploration, detection and measurement in all phases of engineering survey, design, construction and operation.

1.0.3 At present, the applied geophysical methods mainly include electrical survey, ground penetrating radar, seismic exploration, elasticity wave testing, computerized tomography, sonic echo exploration, radioactivity survey, comprehensive logging, etc. Applications of all methods are described in Annex A. In practical applications, based on geophysical characteristics (see Annex B for physical property parameters of common media), features of different methods and field working conditions, one or several appropriate methods are reasonably chosen.

1.0.4 The procedure for engineering geophysical exploration of water resources and hydropower should cover reception of a task, related data collection, field reconnaissance, design of working plan, tests, fieldwork, data check and assessment, data analysis and processing, data interpretation, technical report writing and submission, etc.

1.0.5 The geophysical exploration technical framework shall be established according to the requirements of exploration task, through gathering and analysis of related information about topography, geology, geophysics, meteorology, hydrology and transportation condition, and combined with the actual field reconnaissance. The technical framework shall primarily cover task and objective, survey area and workload, field geological and geophysical characteristics, methods and techniques, personnel and equipment configuration, operating period and scheduling, expected results, etc.

1.0.6 The instrument and its major transducers that are used in geophysical operations shall be tested and permitted by the department concerned in the specified testing period and meet the technical indices as required by this code.

1.0.7 Geophysical engineers shall collect relevant data in a timely manner to support the arrangement of the geophysical operations and interpretation of data.

1.0.8 The standards referenced in this code mainly include:

GB 6722 *Safety Regulations for Blasting*;

GB/T 14583 *Norm for the Measurement of Environmental Terrestrial Gamma-radiation Dose Rate*;

GB/T 14582 *Standard Methods for Radon Measurement in Environmental Air*;

GB/T 16356 *Standard for Controlling Radon and Its Daughters in Underground Space*;

GB 18871 *Basic Standards for Protection Against Ionizing Radiation and for the Safety of Radiation Sources*;

GB/T 50287 *Code for Water Resources and Hydropower Engineering Geological Investigation*;

GB/T 50027 *Standard for Hydrogeological Investigation of Water-supply*;

SL 275 *Field Test Methods for Nuclear Density-moisture*

Gauges;

SL 73.3 *Standard for Engineering Drawing of Water Resources and Hydropower (Survey and Investigation)*;

JGJ/T 23 *Technical Specification for Inspection of Concrete Compressive Strength by Ultrasonic Rebound Method*;

JGJ 106 *Technical Code for Testing of Building Foundation Piles*;

CECS 21 *Technical Specification for Inspection of Concrete Defects by Ultrasonic Method*.

1.0.9 Not only the requirements stipulated in this code, but also those in the current relevant ones of the nation shall be complied with.

2 Terms and Notations

2.1 Terms

2.1.1 Electrical sounding

An electrical survey method used to investigate the distribution of subsurface media according to the electrical contrast between the target and the host material, based on gradually deeper investigation depth with expanded electrode spacing at a given station where the shallow-to-deep resistivity changes are vertically measured.

2.1.2 Electrical profiling

An electrical survey method used to investigate the subsurface lateral earth resistivity changes at a given depth along the survey line with the electrode spacing of some array remaining fixed according to the electrical contrast between the target and the host material.

2.1.3 Resistivity imaging

A multi-electrode profiling technique that records more than hundreds of subsurface data points (so called high density) which are used to reconstruct a colored resistivity cross-section of the earth. It's an essential combination of the electrical sounding and electrical profiling.

2.1.4 Induced polarization

An electrical survey method used to investigate the distribution features of subsurface media according to the difference of induced polarization effect between the object and its surrounding medium.

2.1.5 Self-potential method

An electrical survey method used to understand the hydrological and engineering geology in terms of measuring the law and behaviors of self-potential which arises from electrochemical action in the subsurface media, particulate filtration in the ground water, salt dispersion and adsorption, etc.

2.1.6 Mise-a-la-masse method

An electrical survey method used to investigate the distribution features of an object by powering the surveyed object to increase the potential difference between the object and its surrounding medium to get a charging effect.

2.1.7 Controlled source audio frequency magnetotellurics (CSAMT)

An electrical survey method used to investigate the distribution information of medium resistivity at varied depths and distribution features of an object by measuring the frequency responses of a ground electromagnetic field generated by audio-electromagnetic signals from artificially controlled sources. This method is based on the principle of the varying penetration depths of electromagnetic waves at different frequencies.

2.1.8 Transient electromagnetic method (TEM)

An electrical survey method used to investigate the subsurface medium features by using the non-grounded loop or grounded-wire dipole to radiate pulse electromagnetic waves into the subsurface so as to measure the secondary electromagnetic field generated by the subsurface eddy current induced from the pulse electromagnetic field.

2.1.9 Ground penetrating radar (GPR)

A geophysical exploration method used to investigate the distribution of an object by utilizing a transmitting radar antenna to radiate high frequency pulse electromagnetic waves into the subsurface and a receiver antenna to receive the reflected electromagnetic waves of the object.

2.1.10 Shallow seismic refraction

A seismic exploration method (shallow seismic refracted wave for short) used to investigate the shallow layers or structures that have different wave velocities by using the refraction principle of seismic waves.

2.1.11 Shallow seismic reflection

A seismic exploration method (shallow seismic reflected wave for short) used to investigate shallow layers or structures with different wave impedance by using the reflection principle of seismic waves.

2.1.12 Rayleigh wave method

A seismic exploration method used to identify the layered medium by using the geometrical frequency dispersion properties of Rayleigh waves in stratified medium. It can be divided into steady state and transient state by shock excitation.

2.1.13 Vertical reflection method

An elastic wave survey method used to investigate an object according to the elastic wave reflection principle by moving a pair of source-receiver with the minimal equal offset along a survey line. The data analysis and interpretation are based on the variation features of phase, amplitude and frequency of the reflection information.

2.1.14 Elasticity wave testing

A method used to measure the velocity of rock and soil and concrete or to detect the flaw of concrete by using the kinematics and dynamics features of elasticity wave.

2.1.15 Computerized tomography (CT)

A method used to reconstruct wave velocity or energy absorption images of media by using the transmitted elastic waves or electromagnetic waves.

2.1.16 Sonic echo exploration

A method used specifically to investigate the underwater topography or stratify the underwater layers, by using the sound wave reflection principle. It is also called sub-bottom profile.

2.1.17 Radioactivity survey

A method used for exploration by using the natural or artificial radioactivity features of media.

2.1.18 Isotope tracer technique

A radioactivity surveying method used to mark the groundwater flow of a natural flow field or an artificial flow field and to test the direction of groundwater flow and its velocity following the tracing or dilution principle by using the artificial radioactivity isotopes (^{131}I, ^{182}I, ^{85}Br, etc.).

2.1.19 Comprehensive logging

A comprehensive survey method used to measure the physical properties of media around the borehole by using two or more geophysical logging techniques.

2.1.20 Environmental radioactivity detection

A method used to measure the radioactivity strength in engineering and living environment as well as natural building materials by using the radioactivity measurement method.

2.2 Symbols

2.2.1 The statistical parameters mainly include the following:

M—total relative error of mean square

m—relative error of mean square

K—range coefficient

Δ—absolute error

$\bar{\Delta}$—mean absolute error

δ—relative error

$\bar{\delta}$—mean relative error

2.2.2 The electromagnetic parameters mainly include the following:

I—current intensity

U—potential

σ—conductivity

μ—magnetic conductivity

ε—dielectric constant

ρ—resistivity

α—attenuation coefficient

β—coefficient of medium absorption of electromagnetic wave energy

D_s—apparent attenuation coefficient

E_x、E_y—component of electrical field

H_x、H_y—component of magnetic field

ΔU—potential difference

ΔU_g—interference potential difference

J_s—apparent excitation ratio

η_s—apparent polarizability

S_τ—longitudinal conductance

ρ_s—apparent resistivity

ΔU_1—potential difference of primary field

ΔU_2—potential difference of secondary field

$S_{0.5}$—half attenuation time

2.2.3 The physical mechanical parameters mainly include the following:

A—amplitude

K—reflection coefficient, calibration factor

V—velocity of wave

μ—Poisson's ratio
f—frequency of wave
ρ—density
λ—wavelength
E_d—dynamic modulus of elasticity
E_s—static modulus of elasticity
G_d—dynamic shear modulus
f_c—compressive strength of concrete
V_P—velocity of P-wave
V_S—velocity of S-wave
V_R—velocity of Rayleigh wave
V_b—interface velocity
V_a—mean velocity
V_f—horizontal flow velocity
V_v—vertical flow velocity
V_{ef}—effective flow velocity
V_{pr}—P-wave velocity of intact rock
$V^{//}$—velocity parallel to orientation of rock structure plane
$V^{\perp}$—velocity perpendicular to orientation of rock structure plane
K_0—unit elastic resistance coefficient

2.2.4 The geometrical, distance and azimuth parameters mainly include the following:

AB—spacing of source electrode

MN—spacing of measuring electrode

OC—infinite spacing of source electrode at the measuring point O

OA—positive spacing of source electrode at the measuring point O

OB—negative spacing of source electrode at the measuring point O

K—electrical array factor

L—side length of transient electromagnetic frame

R— the maximum elongation distance of self-potential circles of equal potential

d—distance, spacing between the transmitter and receiver, width of fault

H—elevation, buried depth, hole depth, thickness

δ—skin depth of electromagnetic wave

α—azimuth

x—spacing between stations

ΔR—displacement increment of self-potential adjacent circles of equal potential

X_0—offset, spacing between the source and geophone

S_R—loop area of transient electromagnetism

2.2.5 The time parameters mainly include the following:

T—circle

t—time

ω—angular frequency

τ_s—apparent time constant

Δt—time interval

2.2.6 Other parameters mainly include the following:

n—porosity

t—temperature

η—coefficient of anisotropy

K_w—coefficient of weathering

K_v—integrality index

3 Geophysical Methods and Techniques

3.1 General

3.1.1 An overall understanding and analysis shall be made of the survey areas in terms of topography, geology and geophysical conditions as well as the technical results of the previous work before fieldwork.

3.1.2 The check and use of instruments shall meet the following requirements:

　　1 Periodical testing of equipment shall be undertaken according to the required testing period of instrument and technical indices, and the testing results of each time shall be recorded.

　　2 The equipment shall be checked prior to fieldwork. For the same type of equipment used in an area, uniformity comparison shall be conducted at the same station by using the same survey configuration and same survey mode.

　　3 If an instrument is out of order in the process of fieldwork, it shall not be used until the fault is removed and the normal work is confirmed.

　　4 Instrument shall be checked after fieldwork is finished.

3.1.3 The layout of survey grids and lines shall meet the following requirements:

　　1 The layout of survey grids shall be determined on the basis of comprehensively considering such factors as the operation requirement, investigation method, dimension of the survey object and its buried depth. The survey grids and working scales shall be so selected that the surveyed object can be reflected and its location and geometry can be clearly marked on the map.

2 The survey lines should be set out at the terrain with small topographic relief and relatively homogeneous surface media and their direction should be perpendicular to layer or structure strike and the strike of a principal surveyed object; the survey lines shall be in compliance with the ones of geologic exploration and other geophysical methods and be far from noise source.

3 When a major anomaly is found in the vicinity of the boundary of a survey area, the survey lines shall be duly extended beyond the survey area to trace the anomaly.

4 The survey lines shall be duly added in the geostructurally complicated area and supplementary survey lines shall be placed in between the principal survey lines.

5 In mountainous areas, survey lines should be set out along the contour line or slope; if there is a small topographical relief, a long survey line may be set out along the hillside of similar slope; if there is a large topographical relief especially along the mountain ridge or on both sides of a valley, the survey line shall be divided into several short segments.

3.1.4 The test work shall meet the following requirements:

1 Before a measurement, a testing program shall be formulated in accordance with the operation requirement and geological and physical property conditions in the survey area, and the testing results may be used as part of the working results.

2 The test work shall follow the principle of "from known to unknown" and "from simple to complex". The testing sections should be representative and located on the survey lines of geophysical operations and go through the borehole if any.

3 Appropriate instrument parameters and specifications

shall be chosen according to the test results.

3.1.5 The observations, repeated observations and check observations shall meet the following requirements:

  1 Observation of excitation and receiving signals shall be made at the time when the background is relatively quiet and the signals are relatively stable.

  2 At the terminal of a survey line, the sudden change point and distortion segment of a curve, repeated observations shall be made if the instrument parameters or observing conditions are changed. The average relative errors of the repeated observations shall be smaller than 5%.

  3 The check observation workload in a survey area or survey lines shall be not less than 5% of the total workload in the survey area or survey lines.

  4 The check points shall be uniformly distributed within the whole survey area and placed specifically for the anomalous sections, questionable points and sudden change points.

  5 The survey work shall be repeated if the error of a check observation in a survey area or survey lines is greater than the requirement in this standard.

  6 The operator shall check each record on site and locate the causes if the record is not up to the requirement and make timely resurvey.

3.1.6 The records of geophysical exploration shall meet the following requirements:

  1 The records of geophysical exploration include equipment inspection, check and maintenance, original record, repeated check record, self-checking record, survey record, result review record, and user feedback record, etc.

  2 The original records include field log book (name of

project, survey area, survey line or borehole, station number, the name of companies and operators, calibration personnel, description of equipment, type, main specifications of the equipment, observation system, etc.), observed data or record, file number, print record of data, and record of anomalous situations in course of observation.

 3 The records of geophysical exploration shall not be altered, erased or missed. The files of computer acquisition data shall be correctly numbered and their contents shall be complete.

3.1.7 The check and assessment of data shall meet the the following requirements:

 1 Field operation personnel shall self-check all the original records.

 2 The engineer in charge shall organize relevant personnel to make check and assessment of the original records with the sampling check rate greater than 30%.

 3 The raw data shall be assessed as qualification or disqualification. The raw data is judged as disqualification under one of the following situations:

  1) Incomplete records.

  2) Alterations, obliteration and missed pages in the original records.

  3) Disagreement of file names of computer acquisition data with their contents or incomplete contents.

  4) Repeat and check observations are not made as required.

  5) Accuracy of check observations is not up to the requirement.

  6) Equipment in use is disqualified.

  7) The undesirable observing systems and arrays are used.

8) Leak detection is not conducted for the equipment needed or fails to pass the check.

9) All data from the equipment that has no equipment check records or has no regular check, or fails to pass the check.

3.1.8 The processing and interpretation of data shall meet the following requirements:

1 The interpretation and inference of data shall be made by fully considering the geological, design and construction data in the area of geophysical exploration, repeated comparison and analysis of the data, in particular the analysis and summarization of the different anomalous occurrences, which are needed before correct conclusions reached.

2 The indoor and field work shall be done simultaneously with the former guiding the latter. A preliminary classification and interpretation of data shall be duly made on site. If the raw data are found in question or their interpreted conclusions insufficient, necessary field work shall be performed to make a supplement.

3 The data shall be synthetically interpreted with a full consideration of the geological conditions, internal relationship between geological conditions and survey results and potential interferences.

4 The interpreted results shall be expressed by using corresponding terminologies.

3.1.9 The finished drawings shall meet the following requirements:

1 The drawings as required by this code shall be in accordance with the provisions of SL 73.3.

2 The drawings shall include layout map, result plot, interpretation plot, etc.

3 The finished drawings include the profiles or the planes obtained by a single geophysical method or integrated geophysical methods, which can be curves, contour maps or images.

4 The interpretation drawings shall be generated from qualitative and quantitative interpretation of geophysical exploration data and correspond to physical properties.

5 The result and interpretation plots for geophysical exploration shall be plotted on a same plot, with the geophysical result map at the upper part and the interpretation plot at the lower part.

3.2 Electrical Survey

3.2.1 The electrical survey includes electrical sounding, electrical profiling, resistivity imaging, self-potential method, mise-a-la-masse method, induced polarization method, controlled source audio frequency magnetotellurics (CSAMT), and transient electromagnetic method (TEM), etc.

3.2.2 The application conditions shall meet the following requirements:

1 The requirements of the electrical survey methods are as follows:

    1) The surveyed layers of interest shall have a certain dimension and lateral extension relative to the buried depth and array length. The surveyed object shall have a certain dimension relative to the buried depth and array length. The surveyed layer of interest and the adjacent layers or the object and its surrounding media shall have a difference in electrical properties, and the electrical interfaces shall be relevant to the ge-

ological interfaces.
2) Topographical relief is small and grounding condition is good.
3) For the electrode ground contact observation, no shield layer with very high resistance exists above the surveyed layer of interest or the object. For the loop observation, no shield layer with very low resistance exists above the target layer or the object.
4) The layers or geological bodies are stable in electrical property. The extension and magnitude of anomalies can be observed and tracked.
5) No strong Industrial stray currents or electromagnetic noises exist within the survey area.
6) Water velocity is relatively small for working on water.

2 The requirements of the electrical sounding method used for stratification include:
1) The underground electrical layers are not many and the electrical marker layers are stable, and the surveyed layers shall have a certain dimension and thickness relative to the current electrode spacing, and have a certain lateral extention.
2) The intersection angle between land surface and the underlying bedrock's beddings or the surveyed layers of interest shall be less than 20°.
3) The resistivity data of a certain number of intermediate layers are available.

3 For the electrical profiling, the intersection angle between the surveyed geological interface or structural planes and the land surface shall be greater than 30°.

4 The requirements of self-potential method used for investigating the seepage field include:

    1) The seepage layer is large in pressure difference and the groundwater is low in salinity. The bedrocks are micro porous in texture and can generate a strong natural electrical field.

    2) The water-saturated seepage layers are not deeply buried, and their underlying and overlying bedrocks are high in resistivity.

5 When the mise-a-la-masse method is used to measure the groundwater velocity and flow direction, the boreholes shall be available in the survey area with a depth below the groundwater table by a certain margin and their metal casing above the groundwater table. The aquifer should be located in a depth of less than 50m with the groundwater velocity greater than 1m/d, and the resistivity of surrounding medium three times as much as the water's resistivity. When the mise-a-la-masse method is used to investigate the geological bodies of low resistivity, the geological bodies shall be 10 times greater than the adjoining rocks in conductivity with their dimension matching their buried depth that should be less than 25m.

6 When the induced polarization method is used to investigate groundwater, obvious electrochemical reaction and electrical charge effect shall exist in the form of ion exchange at the solid/liquid interface. And in the survey area there exist no or few metal minerals, coal seams, graphite, carbonized rocks, and the like that have strong electrochemical reactions.

7 The survey area for the controlled source audio frequency magnetotellurics (CSAMT) shall conform to the requirement for the chosen field sources and the surveyed layer of interest or object

shall be located below the blind zone. The electromagnetic noises and other interferences shall be small.

8 The transient electromagnetic method is suitable for the survey areas that have no condition for deploying the array of electrodes such as desert, gobi, bare rock, frozen soil, etc. Such obstacles as brambles, woods, and steep slope shall not be exist at the survey lines and stations for easily deploying the line frames and external interferences of electromagnetic noises shall be small.

3.2.3 The instrument and devices shall meet the following requirements:

1 Multifunction direct current resistivity instruments should be chosen for the electrical sounding, electrical profiling, self-potential method, mise-a-la-masse method and induced polarization method. The instrument shall have such functions as direct measurement, display, and storage as well as the compensation for the spontaneous potential, drift and electrode polarization, and can measure multiple parameters such as primary field potential, spontaneous potential, supply current, apparent resistivity, integrated IP parameters (apparent polarizability, half attenuation time, and degree of attenuation). The main specifications shall meet the following requirements:

    1) Resolution of observing voltage is 0.01mV.
    2) Resolution of observing current is 0.01mA.
    3) The maximum compensation range is ± 1V.
    4) Input impedance is not less than 8MΩ.
    5) The maximum supply voltage is not less than 900V.
    6) The maximum supply current is not less than 3A.

2 The requirements for the instrument of the CSATM are as follows:

1) The signal frequency range, signal stability and field intensity of controlled sources shall meet the requirements for the survey conditions.
2) The receiving instrument shall have such functions as more than two electrical and magnetic reception channels, digital acquisition, program-control gain, automatic analysis of sounding signals, curve display and evaluation, automatic adjustment of sampling rate and sampling length depending on signal frequency.
3) The receiving electrode is an unpolarizable electrode.

3 For the transient electromagnetic method, the used instrument shall be endowed with multichannels, adjustable sampling rate and length, and signal stack. Its main technical parameters shall meet the following requirements:

1) Transmitting voltage is 12V to 400V.
2) The transmitting base frequency is classified within the range of 2.5Hz to 225Hz.
3) Band width is 10Hz to 7.5kHz.
4) Range of time window is 0.05ms to 160ms.
5) Sensitivity of channel is $0.5\mu V$.
6) Transmitting current is not less than 5A.
7) The number of testing channel is not less than 12.
8) Dynamical range is not less than 140dB.
9) Equivalent input noise is not greater than $1\mu V$.
10) Interference suppression on power frequency is not less than 60dB.

3.2.4 The survey layout shall meet the following requirements in addition to the provisions of 3.1.3 in this code:

1 Coordinate measurement shall all be made for stations of

electrical sounding, sounding stations of CSAMT, base stations beyond the self-potential cross-section, charging points of mise-a-la-masse method, major anomalous points, terminal and turning points of the survey line.

2　The station spacing of the electrical sounding should be 1cm to 3cm on the result plot with the corresponding precision, and the line spacing be 1 to 3 times the station spacing.

3　The requirements for layout of the survey grid of the electrical profiling are:

1) Several paralleled survey lines shall be deployed perpendicularly to the strike of geological structure and lithological boundary for tracing the trend.

2) The survey lines that go through local anomalies shall not be less than 2 and the anomalous points that reflect a given object shall not be less than 3 on each survey line.

3) The line and station spacing may be determined according to the operational requirement, the dimension and buried depth H of the surveyed object. The station spacing should be $H/3$ to H and the line spacing be 2 to 5 times the station spacing.

4) In case of using planar contour map to show the anisotropy of geological body, the station spacing and line spacing should be identical.

4　For the resistivity imaging, the overlapping length between the station spacing and survey line shall be determined by the type of array, the number of electrode, investigation depth, and survey precision, etc.

5　The self-potential method shall be selected for an area with flat topography, wet terrain surface and stable electric fields, and far from the surface runoff. The survey lines may be set out as grid pattern within which a base station (i.e., assumed zero potential point) shall be set. When the survey area is large, several base stations and sub-base stations should be set.

   6 When the mise-a-la-masse method is used to investigate groundwater velocity and flow direction, 8 or 12 spokewise survey lines shall be evenly set out with a borehole as the center, and the direction error not greater than ±5°. When the mise-a-la-masse method is used to investigate the geological body of low resistivity, the geological body shall be taken as the center, and the station spacing should be less than half the buried depth of the surveyed object and the survey line spacing be 2 to 5 times the station spacing, and over 3 profiles shall cross the geological body.

   7 The survey lines for the induced polarization method should be set out at the aquifers or structural locations that are determined by other geophysical methods.

   8 The station spacing for the CSAMT should be 5m to 50m and the line spacing be 1 to 5 times the station spacing.

   9 The layout of the survey grid for the transient electromagnetic method shall take into consideration the size of the line frame and its layout requirement. The line spacing should be 1 to 2 times the side length L of the frame and the station spacing may be as long as L, $L/2$ or $L/4$.

3.2.5 The electrical leakage detection shall meet the following requirements:

   1 The insulance resistivity between the tool housing and electrode shall not be less than 300MΩ and the insulance resistivity of wire not less than 2MΩ/km.

2 Stop measurement when electrical leakage is detected. Repeated measurement of those likely affected points after the electrical leakage is eliminated.

   3 The electrical leakage detection shall be made in such situations as the infinite current electrode as work starts or ends, every other 20 stations in normal condition, moving to a new survey station and completing an operation, the maximum current electrode spacing of the electrical sounding, and the distortion point of observed data.

3.2.6 The fieldwork of the electrical sounding shall meet the following requirements:

   1 The test work includes selection of array type, optimal electrode spacing, optimal supply current, power-on time, station spacing, the azimuth of electrode expansion, and resistivity of rock and soil.

   2 The requirements for selecting arrays are:

   1) Selectable configuration may be schlumberger and bi-directional pole-dipole arrays, dipole and differential arrays, and combinations of more than 2 arrays.

   2) When the surveyed strata has multiple layers of resistivity and enough open space is available for expansion electrodes at both ends of the survey line, the symmetrical and bi-directional pole-dipole arrays should be chosen. When the strata has few resistivity layers with significant electrical contrasts and no enough open space is available for expansion electrodes, the pole-dipole arrays should be chosen.

   3) For the stratified and local unfavorable geological body, the symmetrical and pole-dipole arrays may be selected. For the non-horizontal structural zones and

lithological interfaces, the bi-directional pole-dipole and differential arrays may be selected. For the electrical parameters of rock and soil, the symmetrical array may be selected. For the shallow non-homogeneous geological body, the dipole sounding may be selected.

 4) The combination set of 2 different arrays shall be tested before being used in the area with known geology, and shall not be applied before meeting the task requirement.

3 The requirements for selecting electrode spacings are:

 1) AB, OA or OB shall be in uniform distribution in a log-log coordinate system and the ratio between adjacent electrode spacings should be 1.2 to 1.8. For detailed survey of shallow layers or resistivity parameter measurement, the electrode spacings shall be increased in arithmetical progression.

 2) The resistivity of the first layer shall be measured with the least current electrode spacing and $AB/2$ should be 1.5m. There shall be not less than 3 points with the maximum current electrode spacing AB after the "turning points" reflect the ascent or drop-down curve of the marker layer in the electrical sounding curve.

 3) OC of the pole-dipole or bi-directional pole-dipole sounding shall be located at the MN perpendicular bisector and 5 times greater than the maximum OA or OB. When C electrode is in line with the direction of an array, OC shall be 20 times greater than OA or OB and the effect error of the C electrode on the measured apparent resistivity shall be kept less than 2%.

4) Ratios between MN and either of AB, OA and OB shall be 1/3 to 1/30.

4 The requirements for field arrangement of electrodes are:

   1) The copper electrode shall be used as the measuring electrode and the copper, steel or iron electrodes may be used as the current electrode, and the lead electrodes should be used on water or ice.
   2) The array direction of the electrical sounding shall make the topography have the smallest effect on the measured data, and in case of high voltage line, the array direction shall be perpendicular to it.
   3) The deviation of electrode location shall be less than 1% of the electrode spacing in the electrode expansion direction, and less than 5% of the electrode spacing when perpendicular to electrode expansion direction.
   4) When sounding on water or riverbed, the electrodes shall be arranged on water surface or river bottom. For electrode array on water surface, the electrodes shall be submerged in water and the water depth and coordinates shall be measured at the station position.

5 Controlled cross-shaped or ring-shaped electrical sounding should be uniformly arranged within the survey area with its number not less than 3% of the total electrical sounding stations. No less than 3% of bi-directional pole-dipole sounding should be made for the sounding with pole-dipole arrays.

6 The observations, repeated observations and check observations shall meet the following requirements in addition to the provisions of 3.1.5 in this code:

   1) The power-on time shall be larger than 1s for manual

observation, and greater than 0.5s for the automatic observation.

2) When the ΔU data for electrode spacing are unstable at a sounding station, ΔU is less than 3mV or I is less than 3mA, observations shall be repeated more than 3 times.

3) The supply voltage shall be changed or the electrode grounding conditions shall be improved for repeated and check observations.

4) When observation, repeated observation or check observation are made at one station (a given electrode spacing at the same sounding station), the range factor K may be calculated according to Equation (C.1.7) of Annex C in this code. The data, when K is greater than $\sqrt{n-1} \times 4\%$ (n is the number of the calculated apparent resistivity values), may be rejected, and the mean values of other data may be taken as the final measured data of the station, but the rejected data shall be less than 1/3 of the total observations at the station.

7 The check and assessment of data shall meet the following requirements in addition to the provisions of 3.1.7 in this code:

1) The relative error δ and the relative error of mean square m of the single sounding station shall be calculated according to equations of Annex C.1 in this code. The relative error of the total mean square M is calculated according to equation of Annex C.1 in this code, for one sounding station, one cross-section or one survey area.

2) The data of a sounding station shall be assessed as un-

qualified if one of the following 5 situations occurs: ① 3 adjacent electrode spacings with $\delta > 2.5\%$; ② over 30% of the check electrode spacings with $\delta > 3.5\%$; ③ over 5% of the check electrode spacings with $\delta > 7\%$; ④ over 1% of the check electrode spacings with $\delta > 10.5\%$; ⑤ $m > 3.5\%$.

3) The data of a cross-section or survey area shall be assessed as unqualified if one of the following 3 situations occurs: ① over 30% of the total check stations are unqualified in one cross-section or one survey area; ② $m > 3.5\%$ for all of the checked sounding stations (the unqualified sounding stations included); ③ $M > 3.5\%$.

3.2.7 The fieldwork of the electrical profiling shall meet the following requirements:

1 The requirements for selection of arrays are:

1) Selectable configurations include such arrays as bi-directional pole-dipole, pole-dipole, symmetrical, bipole, or dipole and differential arrays, as well as combination of more than 2 above-mentioned arrays.

2) The bi-directional pole-dipole, pole-dipole, bipole and differential arrays may be chosen for investigating nonhorizontal structural belt, lithological interface and karst. The symmetrical and dipole arrays may be used to investigate the local unfavorable geological bodies. The dipole array may be used to investigate the shallow nonhomogeneous geological bodies.

2 The requirements for determining electrode spacings are:

1) The current electrode spacing should be 3 to 5 times the buried depth of the surveyed object.

2) When the effect of electrical heterogeneity of a surface layer is large, MN should be 1 to 2 times the station spacing and not greater than $AB/3$.

3) The infinite electrodes of bi-directional pole-dipole, pole-dipole and bipole arrays shall meet the relevant provisions of item 3 in 3.2.6 in this code.

4) For the same cross-section, the same array with two different spacings may be used to make investigation at varying depths, but the ratio of the electrode spacings between two arrays should be over 1.5 with their stations superposed.

3 Repeated observations and check observations shall meet the following requirements in addition to the relevant provisions of 3.1.5 and 6 in 3.2.6 of this code:

1) For observation with multiple electrode spacings, all the electrode spacings for the checked stations shall be checked.

2) For the electrical profiling, every 10 observation stations shall have one station for repeated observation.

4 In addition to the provisions of 3.1.7 in this code for check and assessment of data, the data for a single survey line shall meet the provisions of 7 in 3.2.6 in this code.

3.2.8 The fieldwork of the resistivity imaging shall meet the following requirements:

1 The requirements for selecting arrays are:

1) Selectable configurations include the symmetrical, bi-directional pole-dipole, pole-dipole, dipole, differential and middle gradient arrays.

2) For stratification prospecting, the symmetrical and pole-dipole arrays should be chosen. For local unfavorable

geological body prospecting, the symmetrical array should be chosen. For nonhorizontal structural zone and lithological interface prospecting, the bi-directional pole-dipole, pole-dipole, bipole and differential arrays should be chosen. For the shallow nonhomogeneous geological bodies, the dipole array should be chosen.

2 The requirements for selecting electrode spacings are:

1) The basic and observing electrode spacings should be equal to the station spacing.

2) The infinite electrodes with bi-directional pole-dipole, pole-dipole and dipole arrays shall meet the relevant provisions of 3 in 3.2.6 in this code.

3) The current electrode spacing corresponding to the planned deepest layer shall be more than 3 times the investigation depth as required.

3 The onsite electrode arrangement shall meet the following requirements in addition to the provisions of 4 in 3.2.6 in this code:

1) At the terminal of a survey line, the investigation shall be made within the effective range determined by the chosen arrays and electrode arrangement conditions.

2) The electrodes of the same array shall be in line arrangement.

3) Before measurement, check grounding conditions of all the electrodes and ensure the correct connection order.

4 The repeated observation and check observation shall meet the following requirements in addition to the provisions of 3.1.5 in this code:

1) Repeat observation may be conducted by choosing two layers or two rows after each array prospecting is finished.
2) Check observation may be conducted by spot checking method.

5 In addition to the provisions of 3.1.7 in this code for check and assessment of data, the data for the single spread shall meet the relevant provisions of 6 in 3.2.6 in this code.

3.2.9 The fieldwork of the self-potential method shall meet the following requirements:

1 Potential observation should be used and gradient observation may also be used. For prospecting groundwater seepage direction, the ring-shaped observation should be added.

2 An unpolarizable electrode shall be used and the potential difference shall be duly observed before and after an observation. The absolute value of the electrode potential difference shall be less than 2mV at the beginning of the operation and less than 5mV at the end of the operation.

3 The observations, repeated observations and check observations shall meet the following requirements in addition to the provisions of 3.1.5 in this code:

1) The joint observation shall be made between the base stations.
2) The self-potential compensation shall be turned off during observation and the observed values shall be marked with the positive/negative of potentials.
3) When the survey line is relatively long or stray current has a big influence, observation shall be undertaken in segment, and 3 stations shall be repeated at the connection.

4) One repeated observation shall be made for every 10 observation stations.

4 The check and assessment of data shall meet the following requirements in addition to the provisions of 3.1.7 in this code:

1) The absolute error Δ for the single station and the observed mean absolute error $\overline{\Delta}$ for a survey line or a survey area shall be calculated according to Equation in Annex C.1 in this code.
2) The absolute error Δ for the check observation of the single station shall be less than 3mV.
3) The mean absolute error $\overline{\Delta}$ of potential observation for a survey area shall be less than 5mV. The mean absolute error $\overline{\Delta}$ for gradient observation shall be less than 3mV.

3.2.10 The fieldwork of the mise-a-la-masse method shall meet the following requirements:

1 The direct-current may be adopted and a multi-function geoelectric equipment may be chosen. In case of interference, the low-frequency alternating-current may also be adopted and the apparatus with frequency-selection function may be chosen.

2 The pole-dipole array shall be used with the unpolarizable electrode used as the measuring electrode and the potential method or gradient method shall be used for observations.

3 The requirements for prospecting velocity and flow direction of groundwater are:

1) The distance between the borehole and the infinite B electrode shall be more than 15 times the depth of aquifer to be observed, with good grounding condition. The charging electrode A shall be placed at

the middle of the aquifer in the borehole, strung together with salt gauze bags. The distance between the observing electrode N and the borehole shall be 1.5 times the depth of the charging electrode A which shall be fixed in the borehole at some point upstream of the predicted water flow direction.

2) The electrode M shall be moved separately along each survey line to find the equipotential point.

3) The normal equipotential lines shall be measured once before salt is dropped.

4) The equipotential lines shall be measured once immediately after salt is dropped to modify the electrode N's position, based on which the re-measured equipotential lines are used as the new reference point. Then, the equipotential lines are measured once at regular intervals (about 20min to 30min, depending on velocity of the aquifer), and the number of the measured equipotential lines after dropping salt shall be more than 3 lines.

5) For measurement of the equipotential lines, the measurement time shall be recorded and the distance between each equipotential point and the borehole shall be measured.

6) The fixed observing electrode N_1 and N_2 with different distances may be used to make the measurement.

4 The requirements for investigating the geometry of a low-resistivity geological body are:

1) The distance between the infinite electrode B and the current electrode A shall be more than 10 times the buried depth of the low-resistivity geological body or the extended length, and the current electrode A shall

be in good contact with the low-resistivity geological body.

    2) When the potential observation method is adopted, the electrode N shall be located at the opposite direction of electrode B, and the distance between electrode and the current electrode A shall be greater than 10 times the buried depth of the low-resistivity geological mass or the extended length; the spacing of the observing electrode shall be 5m to 10m in using the gradient method to take observations.

5 In addition to meet the provisions of 3.2.5 in this code, the electrical leakage detection shall meet the requirement that the resistance of MN circuit is less than $10k\Omega$ and the insulation resistance of the electrode wire is greater than $5M\Omega$.

6 In addition to the provisions of 3.1.5 in this code, the observations, repeated observations and check observations shall meet the following requirements.

    1) The continuous and stable supply current shall be ensured in observation, and may be between 0.5A to 1A.

    2) In the repeated observations and check observations, the relative error δ and the mean square relative error m shall be calculated according to equation in Annex C.1 in this code.

    3) The repeatedly observed δ for the single station shall be less than 5%.

    4) The check-observed m for a survey line shall be less than 7%.

    5) The check-observed m for a survey area shall be less than 7%.

6) The data from the stations with severe interferences may be excluded from the observation quality evaluation for a survey area.

3.2.11 The fieldwork of the induced polarization method shall meet the following requirements:

 1 The 4 electrode array should be chosen. For induced polarization profiling in a large area, the combined parallel array or middle gradient array may be used.

 2 Selection of electrode spacings and onsite electrode arrangement shall meet the following requirements in addition to the provision of 3 and 4 in 3.2.6 in this code:

  1) The maximum current electrode spacing AB for the symmetrical array shall be 3 times greater than the survey depth.

  2) The unpolarizable electrode shall be used in the measuring electrode.

  3) The current wire and measuring wire shall be separated for at least 1m, and the separation distance shall increase in proportion to the wire length.

 3 In addition to the provisions of 3.1.5 and 6 in 3.2.6 in this code, repeated observations and check observations shall meet the following requirements:

  1) The power-on time should be greater than 30s during observation with a stable supply current. The current shall increase in proportion to the current electrode spacing, and ΔU_1 shall be greater than 0.3mV.

  2) The repeated observations shall be made if one of the following situations occurs: ①the potential difference ΔU_2 for secondary field is smaller than 1mV; ②the apparent polarized ratio J_s is greater than or

approximate to the apparent polarizability value η_s; ③the apparent attenuation value D_s is greater than or approximate to 100%.

3) The range coefficient K for repeated observation data shall be less than $\sqrt{n-1} \times 7\%$ (n is the observation frequency). Otherwise the repeated observation shall be added.

4 In addition to the provisions of 3.1.7 in this code, the check and assessment of data shall also meet the following requirements:

1) For the single station or sounding station with the induced polarization method, the requirements for the mean square relative error m for an electrode spacing at the corresponding station are that the apparent polarizability shall be less than 5%, the apparent polarized ratio less than 7%, the apparent attenuation coefficient less than 7% and the half attenuation time less than 7%.

2) If the unqualified sounding stations by check exceed 30%, the data for the survey line or survey area are unqualified.

3) If the unqualified profiling and sounding stations exceed 30% of the total checked stations, the data for the survey line or survey area shall be assessed as unqualified.

3.2.12 The fieldwork of CSAMT shall meet the following requirements:

1 The requirements for choice of field sources and layout are:

1) The electrical dipole or magnetic dipole should be chosen for artificial source. The dipole or magnetic

dipole source shall be chosen based on the buried depth of a surveyed object.

2) The transmitter-receiver distance d (from dipole midpoint to the observation point) and the maximum buried depth of the surveyed object H_{max} should be $5H_{max} \geqslant d \geqslant 3H_{max}$, the length of the electrical dipole should be equal to H_{max}.

3) The electrical dipole shall be arranged parallel to the survey line, with its direction error less than 5°.

4) The current electrode should be placed in the moist soil, and buried in pit (1m to 2m in depth, 1m² to 2m² in area) and joined with multi-layer metal sheets, nets and tin foils, or arranged in ring and parallelly connected to muliple electrodes. If the grounding conditions are poor, brine water may be doused and the grounding resistance should be less than 30Ω.

5) The magnetic dipole shall be placed at a flat and dry area and its axis direction shall be perpendicular to earth with the angular error less than 5°.

2 The on site electrode arrangement shall meet the following requirements in addition to the provisions of 4 in 3.2.6 in this code:

1) In receiving, either two-component observations of E_x and H_y or E_y and H_x, or four-component observations of E_x, H_y, E_y and H_x may be selected and the direction of electrode arrangement shall be in line with the planned electrical component direction, and the direction of the magnetic bar is perpendicular to the one of the electrode spread with their direction errors less than 5°.

2) The unpolarizable electrode shall be used in the

electrical channel and the magnetic probe with the corresponding frequency shall be used in the magnetic channel.

3) The electrolyzing unpolarizable electrode shall be half buried in a pit in the soil that is watered. When high-resistance bare rock occurs at the station, the electrode should be buried with soil that is watered.

3 In addition to the provisions of 3.1.5 in this code, the observations, repeated observations and check observations shall meet the following requirements:

1) The observations of the electrical dipole field sources shall be performed in the far-field region within the 30° fan area on both sides of the electrical dipole AB's midnormal.

2) The observations shall be made in the period when interference background is small.

3) During observation, the connecting wire of the electrode or magnetic bar shall not be hanging, swinging or forming like a coil, and the magnetic bar shall be horizontally placed at the station position and the receiver, operator and magnetic object shall be kept away from the magnetic probe.

4) The electromagnetic fields shall be transmitted and observed from high-frequency to low-frequency with the frequency range matching the investigation depths.

5) After observation is completed for each point or station, the apparent resistivity and phase curve shall be displayed or printed in a timely manner, and the shape between the apparent resistivity curve and phase curve shall be similar for each observation.

6) When the extreme point on the apparent resistivity and phase curve displace at the frequency axis or the curve types changes, a repeated observation shall be made. The shape of the apparent resistivity and phase curve from the repeated observations or check observations at a given station shall be in line with the corresponding magnitude.

7) In case of transferring or changing field sources, more than 3 stations at a given survey line shall be covered, which can be used as a check station.

4 In addition to the provisions of 3.1.7 in this code, the check and assessment of data shall also meet the following requirements:

1) When taking the observations of two-component at a given sounding station, the data at the station are assessed as the qualified if the next two situations are all satisfied: ① over 75% are the points with the standard deviation less than 40% in the apparent resistivity curve; ②over 75% are the points which do not exceed 45°or 135° in the phase curve.

2) When taking the observations of four-component at the same sounding station, the data at a given station are assessed as the qualified if both of the next two situations are satisfied: ①over 70% are the points with the standard deviation less than 40% in the apparent resistivity curve; ②over 70% are the points which do not exceed 45°or 135° in the phase curve.

3) If the unqualified sounding stations exceed 30% of the total checked sounding stations, the data for the survey line or survey area are unacceptable.

3.2.13 The fieldwork of the transient electromagnetic method shall meet the following requirements:

  1 The requirements for choice of arrays and parameters are:

    1) The array with the coincident loop and central loop should be chosen for investigating shallow media, the array with the large-fixed loop for deeper investigation depth and the dipole array for investigating steep faults.

    2) The side length of wire frames L for the arrays of the coincident loop, central loop and dipole should be 0.5 to 1.0 time H_{max} of the maximum buried depth of the surveyed object. The side length of the transmitting wire frames for the large-fixed loop array should be chosen within the range of 100m × 200m to 300m × 600m according to the investigation depth. The side length of the receiving wire frames for the central loop, large-fixed loop and dipole arrays, and the spacing of the transmitting and receiving wire frames, are selected in terms of testing.

  2 The wire frame arrangement shall meet not only the provisions of 4 in 3.2.6 in this code, but also the following requirements:

    1) The position error of the array measuring center at the station shall be less than 10% of the side length of the receiving wire frame, and the point position error between the wire frame corner point and the designed wire frame corner point shall be less than 5%.

    2) Such interferences shall be kept away from railway, underground metal pipe, high-voltage line, transformer, power transmission line, etc.

3) When laying a wire frame, the remainder of a wire is not too long and shall be laid in "S" shape at the land surface or away from the survey area.

3 In addition to meeting the provisions of 3.2.5 in this code in leak detection, it is also required that the resistance for the power line per kilometer shall be less than 6Ω and that the wire shall be soft and wear-resistant with its insulation resistance greater than $2M\Omega$.

4 In addition to the provisions of 3.1.5 in this code, the observations, repeated observations and check observations shall meet the following requirements:

1) The time window shall be determined through a field test and the multi-channel observations shall be taken.

2) The field observation values are allowed below the noise level only in the last 3 to 5 channels. Otherwise the relevant cause shall be identified and a repeated observation made by increasing the multiple stacking number. When a transient interference occurs, the observation is not made until the interference is eliminated.

3) When there is a greater error of the repeated observation at a station, the cause shall be identified, and if there is any interference, a point position may be relocated to avoid it for a repeated observation and also a detailed record is made.

4) An infill station shall be made in the vicinity of an anomaly, and if attenuation of a curve is slow, the time frame shall be widened for a repeated observation.

5) Having finished observations at each station, data and

curves shall be checked and if qualified, moving to the next.

5 In addition to the provisions of 3.1.7 in this code, the check and assessment of data shall also meet the following requirements:

 1) The shape and magnitude of the curves from the observations, repeated observations and check observations shall be in uniformity and the total mean square relative error M for each observation channel shall be less than 10%.

 2) The total mean square relative error M by check for a survey line or survey area shall be less than 15%.

3.2.14 In addition to the provisions of 3.1.8 and 3.1.9 in this code, the data processing and interpretation and plots for the electrical sounding method shall also meet the following requirements.

 1 The qualitative and quantitative interpretation of electrical sounding data shall be made:

 2 The requirements for qualitative interpretation are:

  1) The qualitative interpretation shall include determination of the number of electrical layers, the relationship between all electrical layer resistivities, approximate plane position and the nature of the local anomaly, etc.

  2) The analysis of electrical sounding curve types shall be made, based on which the number of layers and the electrical structure of a geoelectric profile are determined.

  3) A comparison analysis shall be made to determine the relationship between the variation in the electrical sounding curve types for one or multiple profiles and

the layer structure, electrical parameter variation and topographical change.

4) The nature and position of an anomaly shall be determined according to the anomalous magnitude, shape and distribution in an apparent resistivity profile, and also analysis of the anomaly correlation with its adjacent profile.

5) When using the bi-directional pole-dipole for sounding, the nature, magnitude and occurrence of an anomaly shall be determined according to the intersections of the curves in an apparent resistivity profile for all electrode spacings, magnitude and range of an anomaly.

6) When identifying a geoelectric profile, an analysis shall be made of the probability of inconsistency between the electrical and geological interfaces, as well as of the lateral electrical variations and the effect of the topographical change on electrical sounding curves.

3   The quantitative interpretation, which is based on the qualitative interpretation, shall be undertaken to calculate thickness of each electrical layer and to determine the position, magnitude, buried depth and occurrence of an anomalous body.

4   The quantitative interpretation of an anomaly shall be made on the basis of the calculated and measured apparent resistivity profile, magnitude, depth, axial line deviation and plane location of the anomaly in the curves of a bi-directional pole-dipole sounding resistivity profile.

5   The quantitative interpretation shall have the following conditions:

1) Completeness of a curve, significant reflection of the

electrical marker layer reflected on the electrode spacing.

2) Significant electrical layering and definition of types in an electrical sounding curve.

3) The electrical sounding curve, by means of error elimination, roundness and correction of distortion, will not affect the interpretation precision.

4) Available electrical parameters as required by the quantitative interpretation.

5) Rough correspondence of an electrical interface to a geological interface.

6 What are used in the quantitative interpretation are: comparison with theoretical curves contained in albums; computer interpretation of whole curve; electrical reflection factor "K profile" method; variety simplified interpretation or empirical methods.

7 A topographic correction may be made to reduce the effect of topographical relief on apparent resistivity in a resistivity profile.

8 The plots may include apparent resistivity profile or distribution map, geologic interpretation profile or plan-view map for geophysical exploration results (inclusive of layering, erosion zoning, poor geological mass distribution, contour line for bedrock top plate, etc.).

3.2.15 In addition to the provisions of 3.1.8 and 3.1.9 in this code, the data processing and interpretation and plots for the electrical profiling method shall also meet the following requirements:

1 The electrical profiling shall be mainly used for the qualitative interpretation, but if the geometry of a surveyed object is simple, the quantitative interpretation may be made.

2 The qualitative interpretation of an anomaly shall be made through analysis of the features of its magnitude, scale and shape in a curve. When there is a single anomaly, the type of an anomalous curve shall be determined, based on which the qualitative/quantitative interpretation can be conducted. The contrast analysis of anomalies for adjacent profiles shall be made.

3 The topographic correction may be used to eliminate spurious anomalies due to the topographical relief.

4 The plots shall include apparent resistivity profile or distribution map, or geologic interpretation plot for geophysical exploration results, etc. The apparent resistivity curve is plotted using different line styles or colors, below the lateral coordinate of which the topographical profiles, positions, sizes and occurrences of inferred object positions are mapped.

3.2.16 In addition to the provisions of 3.1.8 and 3.1.9 in this code, the data processing and interpretation and plots for the resistivity imaging method shall also meet the following requirements:

1 The apparent resistivity profiles of the resistivity imaging shall be mapped for a whole survey line, and the corresponding resistivity images may also be produced using processed and inversed resistivity data.

2 The following techniques may be used in interpretation:
   1) The interpretation shall be conducted based on the distribution, magnitude and scale of anomalies in apparent resistivity profiles or images.
   2) Several sets of data that meet the conditions for the electrical sounding may be chosen to carry out layering inversion and interpretation.

3) The interpretation shall be made by comparison of apparent resistivity profiles or images of different configurations in a given cross-section.

4) The comparison analysis and interpretation may be made by integrating an apparent resistivity profile or images and the data from the known geological interfaces and boreholes.

3 The plots may include apparent resistivity profiles or resistivity images, geologic interpretation profiles of geophysical exploration results or plan-view maps.

3.2.17 In addition to the provisions of 3.1.8 and 3.1.9 in this code, the data processing and interpretation and plots for the self-potential method shall also meet the following requirements:

1 All the measuring potentials for a surveyed area shall be converted to the same total base point and the potential difference correction of all measuring potentials or gradients shall be made.

2 Background values are determined in sections in accordance with geological physical properties and environmental conditions, with the interferences eliminated.

3 The normal field and anomalous field may be discriminated to make correct identification of the useful anomaly and interferential anomaly. The anomalous values shall exceed 3 times the measured mean absolute error $\overline{\Delta}$ and have a definite regularity and distribution limit.

4 The dimension and buried depth of an anomalous body can be identified based on its distribution and magnitude in the profile or spontaneous potential or gradient curve.

5 The "8"-shaped potential map can be used to calculate the direction of groundwater flow.

6 The plots can include spontaneous potential or gradient

profile curves, profile plan-view maps, contour maps, profiles or plan-view maps for geophysical exploration interpretation results of spontaneous potential or gradient.

3.2.18　In addition to the provisions of 3.1.8 and 3.1.9 in this code, the data processing and interpretation and plots for the mise-a-la-masse method shall also meet the following requirements:

　　1　In testing flow velocity and direction of groundwater, the distribution map of charging isopotential lines shall be plotted at the time of every test.

　　2　In investigating a low-resistivity geological mass, what should be mapped includes the profile curves of charging potential or electric potential gradient, interpretation profiles or plan-view maps of geophysical exploration results for the charging potential or electric potential gradient. The interpretation shall meet the following requirements:

　　　　1) The identification of normal field and anomalous field shall be conducted in terms of the provisions 3 of 3.2.17 in this code to analyze the magnitude, range and shape of anomalies for potential or gradient curves and to determine the dimension and buried depth of a low-resistivity geological mass.

　　　　2) An analysis shall be carried out for the effects of variations on anomalies in the inhomogeneous surface layer, topography, layer occurrences, radial flow of surface water and thickness of overburden.

　　　　3) The planar point position is determined based on the maximum value of a potential profile curve or the "zero" value of a gradient profile curve.

3.2.19　In addition to the provisions of 3.1.8 and 3.1.9 in this code, the data processing and interpretation and plots for the induced polarization method shall also meet the following requirements.

1 The profiling data shall be mainly used for qualitative interpretation and the dimension of an anomalous polarized body can be determined according to the anomalous position in a profile. The buried depth and dimension of the anomalous polarized body shall be identified according to resistivity and multiple induced-electrical parameters for sounding data.

2 The division of background values and anomalous values should be dependent on the observed data of ρ_s, η_s, J_s, D_s and $S_{0.5}$ above the known water table or near a dry hole, which serve as the background value and shall meet the provisions 3 of 3.2.17 in this code.

3 The hydrological geology data shall be combined to conduct an analysis of curve features such as the magnitude, range and shape of an induced-electrical anomaly, and of the relationship between the curve features and the known groundwater, presenting inference and interpretations about the polarized body in the unknown terrain or water abundance of groundwater, buried depth, dimension, etc.

4 The plots may include profiles and plan-view maps for ρ_s, η_s, J_s, D_s and $S_{0.5}$, geological interpretation plot for relevant geophysical exploration results, type plot for a sounding curve, distribution map and buried depth diagram for a water bearing layer, etc.

3.2.20 In addition to the provisions of 3.1.8 and 3.1.9 in this code, the data processing and interpretation and plots for the controlled-source audio-magnetotellurics shall also meet the following requirements:

1 The requirements for pre-processing are:
1) Smoothing, interpolation and correction of the acquired data may be conducted.
2) Check and remove distortion points without automatic smoothing of data and deleting frequency points freely, and conduct the correction of the frequency points with severe distortions at the beginning and the end of a curve with reference to the adjacent stations.
3) The arithmetic average processing can be made of the overlapped points with a relative error of less than 10%, or no processing can be made. The translational processing can be made of the overlapped points with a relative error of greater than 10%, but uniform curve shape.
4) The parameters and optimal static displacement correction shall be chosen in accordance with the known geological data, contour map of an original profile and topographical relief.

2 The requirements for interpretation of data are:
1) The qualitative and quantitative interpretations shall be combined.
2) The qualitative interpretation may be made in terms of the sounding curve types, distribution of anomalous resistivity values in a forward resistivity profile to estimate and identify the electrical models, anomaly properties and distribution.
3) The multi-component sounding curve types and anomalies of the resistivity profile shall be compared in the same four-component measured profile.
4) The comparison of the inversion resistivity profile or

typical sounding curve may be made with the layers indicated from the borehole and poor geological mass, to establish a corresponding relationship between geoelectric models.

    5) The two-dimensional interpretation shall be used in the quantitative interpretation. The known borehole target depth should be used to make a correction of the anomaly depth.

  3 The plots may include resistivity profile and geologic interpretation profile of geophysical exploration results.

3.2.21 In addition to the provisions of 3.1.8 and 3.1.9 in this code, the data processing and interpretation and plots for the transient electromagnetic method shall also meet the following requirements:

  1 The filtering processing of data may be made.

  2 The processing software shall be used to compute and map apparent resistivity and apparent longitudinal conductance profiles.

  3 The division of ambient field and anomalous field shall be made in terms of transient electromagnetic response time characteristics and profile curve types to construct geo-electrical models and identify anomalies.

  4 The corresponding relationship between the apparent resistivity profile in a survey area and its anomaly natures, buried depth and range may be established by borehole data, based on which the interpretation of other profiles in the survey area is undertaken.

  5 The relevant interpretation profiles of geological results or plan-view maps shall be plotted.

3.3 Ground Penetrating Radar

3.3.1 The techniques of profiling, wide angles, loops, trans-illumination, multi-antennas and cross-hole radars may be used in the ground penetrating radar.

3.3.2 The application conditions shall meet the following requirements:

1 There shall be a distinct difference of dielectric constants between the surveyed object and its surrounding media, stable electrical properties and significant electromagnetic reflections.

2 The size of target shall have a certain scale in comparison with its depth and the depth should not be very deep, its thickness in the direction of an investigation antenna dipole axis shall be greater than 1/4 of the effective wave length of electromagnetic waves used in its surrounding media, and its length in the direction of an investigation antenna dipole spread shall be greater than 1/4 of the diameter of the first Fresnel zone of electromagnetic waves used in its surrounding media. To discriminate the two adjacent horizontal surveyed object, their minimum horizon distance shall be greater than the diameter of the first Fresnel zone.

3 The antennas shall pass over a relatively gentle topographic surface at a survey line without any obstacles, with ease of moving an antenna.

4 The objects or layers of interest, which are located below the shield layer with extremely high conductance, can not be investigated.

5 There shall be no large-scale metal structures or high electromagnetic wave interferences from radio transmitting sources, etc.

6 There shall not be metal casing in a borehole in case of a single-hole or cross-hole surveys.

7 In case of a cross-hole (cavity) survey, the object shall be in between two holes (cavities) which are co-planer with their spacing being no greater than the effective penetration distance of radar signals.

3.3.3 The equipment performance shall meet the following requirements:

1 The signal gain control shall have the exponential gain function.

2 A/D converting digit number is not less than 16 bit.

3 In station measurement, there is a multi-stacking function with the stacking number not less than 8 times.

4 In continuous measurements, the scanning rate is not less than 128 lines/s.

3.3.4 The survey grid arrangement shall meet the provisions of 3.1.3 in this code. When conducting the station measurements, the station spacing shall be 0.2m to 1.0m and there shall be more than 3 stations that present anomalies in the anomalous zone.

3.3.5 The settings of equipment parameters shall meet the following requirements:

1 The choice of working frequencies of a main radar antenna shall be dependent on the integrated elements covering investigation task requirement, buried depth of a surveyed object, resolution, media characteristics, and whether antenna sizes meet site conditions or not.

2 The record time window shall be determined by Equation (C.3.3) of Annex C in this code in accordance with the maximum investigation depth and the mean electromagnetic wave velocity of overlying layers.

3 The instrumental signal gain shall keep signal magnitudes not beyond 3/4 of the signal monitoring window, making the signals stable when an antenna stands still.

4 The sampling rate to be chosen shall be 6 to 10 times the center frequency of an antenna.

3.3.6 The choice of an antenna shall meet the following requirements:

1 The 8 MHz to 300MHz frequencies for antennas should be chosen for the surface investigation, and when all the antennas with multiple frequencies can meet the requirements for investigation depths, the one with higher frequencies shall be preferred.

2 The high frequency antenna with its corresponding investigation accuracy should be used to carry out the quality detection of cavity lining, with its frequency ranging from 400MHz to 900MHz.

3 The antenna with 900MHz to 1500MHz should be used to test rebars inside the concrete.

4 The single-hole antenna or cross-hole antenna with one transmitter and one receiver shall be used for a borehole investigation according to the investigation task requirement, and the choice of antenna frequencies shall depend on the investigation range and accuracy.

5 The air coupled antenna should be used for the fast-moving vehicle measurements.

3.3.7 The fieldwork shall meet the following requirements:

1 The metal objects near a survey line shall be removed or kept away from during measurement.

2 The insulating mass shall be used for brace fittings for an antenna. The antenna operator shall not wear any metallic articles and shall keep a relatively fixed distance away from the working antenna.

3 During measuring, the plane of the working antenna shall be parallel to the survey plane and their distance comparatively uniform.

4 For the station measurement, its station spacing shall be less than Nyquist sampling interval; when the survey objects have little variations in the geometry and occurrence, the station spacing may be properly widened and the data shall be collected when the antenna stands still.

5 For the continuous measurement, the contrast test shall be made in advance on station measurements and continuous measurements and the antenna moving speed, which presents the similar results from the continuous measurements and station measurements, shall be chosen. The antenna moving speed shall be kept constant and match the scan rate of instruments.

6 When using the bistatic antenna to make station measurements, the antenna spacing can be adjusted to obtain the highest reflected signals from an object. Twice as much as the critical angle may be used as the opening angle of the receiving antenna and transmitting antenna relative to the surveyed object, and also 1/5 of the maximum depth of the surveyed object as the antenna spacing.

7 In using the dipole antenna, the antenna should be so oriented as to make the direction of polarization of an electrical field parallel to the long axis or trend of a surveyed object. In case of the unknown direction of the long axis of a surveyed object, 2 sets of antennas with orthogonal directions should be used to make separate measurements.

8 Record marks shall be in consistence with stake numbers of a survey line. The interferences from marking signal lines shall be avoided in automatically marking and when using the measuring wheel for marking, a correction shall be made once per 10m.

3.3.8 The transillumination and wide angle techniques may be used to test velocities of electromagnetic waves, which, if the conditions allow, may be calibrated with the object of known depth or estimated by the geometrical scanning method for the tubular object, and also calculated using empirical data or relevant layer parameters.

3.3.9 In addition to the provisions of 3.1.7 in this code, the check and assessment of data shall meet the following requirements:

1 The preliminary editing of radar data shall be made available for check and assessment, whose contents include number of survey line, stake number of mileage, profile depth, etc.

2 Only when there is a substantial agreement in the shape and position of anomalies between the measured images and the original ones can the data be regarded as the qualified.

3.3.10 The data processing shall meet the following requirements:

1 Such processing methods are selectable as deletion of useless channels, horizontal proportional normalization, gain control, topographic correction, frequency filtering, $f-k$ dip filtering, deconvolution, migration location, space filtering, point averaging, etc.

2 The choice of processing methods and steps shall be made according to the fieldwork data quality and interpretation requirement. When the reflected signal is weak and the signal/noise ratio of data is low, the deconvolution and migration location processing should not be done. Before conducting the $f—k$ dip filtering and migration location processing, the useless channels shall be deleted and the horizontal proportion normalization and topographic correction made.

3 The frequency filtering may be used in all phases of data processing to eliminate the interference waves of certain wave band.

4 The $f—k$ dip filtering may be used to cancel the interference waves from an inclined layer, but it shall be ensured beforehand that there are not effective reflected waves of the stratigraphies at the same dip.

5 The deconvolution can be used to suppress multiple reflected waves and its reflected wavelet shall be the minimum phase wavelet.

6 The time migration or depth migration techniques can be used to locate the reflected wave interface of an inclined layer and to converge diffracted waves, and the reliable wave velocities of media shall be used for depth migration processing.

7 The two techniques of the valid channel stacking for space filtering and difference between the channels can be used to allow the anomalies to keep better continuity or independence, promoting the interpretation potential of data and images. The other processing techniques have been used before altering the amplitude features of reflected signals.

8 The point averaging technique for smoothing data can be used to eliminate the high-frequency interferences in signals and the number of points available for calculation shall be an odd number and its maximum value shall be less than the ratio of sampling rate to low-pass frequency.

3.3.11 The data interpretation shall meet both the provisions of 3.1.8 in this code and the following requirements:

  1 The crew sheet and field check shall be used to screen anomalous interferences.

  2 The features such as reflected waveforms and energy intensity on the original image shall be utilized to estimate, identify and screen anomalies.

  3 The event of the strong reflected waves and the strong absorbed waves may be tracked through data processing, or the anomaly width and reflection travel time can be used to calculate the plane-expanded range and buried depth of an anomalous body.

  4 For the transillumination method, the anomalies may be estimated based on such features as having the energy shade or not on transmission images and stacking of secondary waves or not. Or quantitative interpretation based on shade intersection, secondary wave shape, relative location of transmitter and receiver.

3.3.12 The result maps and charts shall meet the provisions of 3.1.9 in this code and also to the following requirements:

  1 The plots shall include radar profiles and radar geological result interpretation profiles.

  2 The distribution cross-section of a survey line shall be mapped at tunnels, cliffs, side walls, etc.

  3 Only the anomaly part may be taken on radar profiles and the grey or chromatographic images may be mapped through continuous measurements. The waveform and radar images mapped through station measurements, and survey line numbers, stake numbers, depth and time shall be marked on

the radar images.

4 On the geological result interpretation profile plotted are the layer boundary, anomaly center, range, expanding orientation, etc.

5 The summary of tables can be used to describe anomalous conditions.

3.4 Seismic Survey

3.4.1 The shallow seismic refraction, shallow seismic reflection and Rayleigh wave methods may be used in seismic exploration.

3.4.2 The application conditions shall meet the following requirements:

1 The methods above are all applicable to the survey in layered media and quasi - layered media.

2 The application conditions for the shallow seismic refraction method are:

1) The wave velocity of the tracked layer shall be greater than the velocities of the overlying layers, between which there are a significant wave velocity difference.

2) The tracked layer shall have a given thickness and the thickness of an intermediate layer shall be greater than the one of its overlying layer.

3) The sum of the apparent dip and the critical angle of refracted waves shall be smaller than 90° for the tracked layer along a survey line.

4) The tracked layer interface shall be less undulated and there is not a penetration phenomenon when refracted waves travel along the interface.

5) There shall be a significant velocity difference at a given magnitude between the surveyed object and its surrounding media.

3   The application conditions for the shallow seismic reflection method are:

　　1) There shall be a significant wave impedance difference between the tracked layer and its adjacent layers.

　　2) The tracked layer shall have a given thickness which is greater than 1/4 of the wavelength of an effective wave.

　　3) The layer interface is flat, on which incident waves can generate quite standard reflected waves.

　　4) There shall be a significant fault throw in the surveyed faults.

4   The application conditions for the Rayleigh wave are:

　　1) There shall be a significant velocity difference between the tracked layer and its adjacent layers and between the surveyed object and its surrounding media.

　　2) The tracked layer shall be lateral homogeneous layered media and the tracked poor geologic body shall be in a given dimension.

　　3) The ground surface shall be relatively flat and the layer interfaces are less undulated to eliminate the effects from the complex topography such as trench, scarp, terrace cliff, etc.

3.4.3   The seismic source and equipments shall meet the following requirements:

1   The requirements for the seismic source are:

　　1) The seismic source may be a hammer, weight drop or explosives, etc.

　　2) The seismic source shall excite the seismic pulse of basic frequency needed for the selected working method, and its energy can be controlled and meet the

requirement for the investigation depth.

3) The blaster shall be on the safe side in performance and have the triggering function of a clock loop.
4) The hammer and weight drop sources shall be convenient to operate and have better repeatability.
5) The clock signal delay time is not larger than 0.5ms.

2 The requirements for seismographs are:
1) The digital seismographs with 12 or 24 channels should be used for shallow layers, characterized by signal enhancement, time-delay, inside-trigger and outside-trigger, pre-amplification, filtering, digital acquisition, etc.
2) The sampling rates are optional and the least sampling interval is not longer than 0.05ms.
3) The recording length, which is optional, is not smaller than 1024 sampling points per channel.
4) A/D conversion accuracy is not smaller than 12 bit.
5) The dynamic range is not smaller than 96dB.
6) The bandwidth is 2Hz to 2000Hz.
7) The internal noises of an amplifier are not bigger than $1\mu V$.

3 The requirements for geophones are:
1) Among all the geophones, the difference in inherent frequency shall be smaller than 10%, the difference in sensitivity smaller than 10% and the phase difference smaller than 1ms.
2) The insulation resistance is not smaller than $10M\Omega$.
3) The hydrophones used in hole and underwater shall have higher waterproof quality.

4 The requirements for consistency of seismic recording

channels are:

    1) The phase difference between one channel and another shall be smaller than 1.5ms.

    2) The amplitude difference between one channel and another shall be smaller than 15%.

3.4.4 The preparation of instrument and equipment shall meet the following requirements:

1 Before fieldwork, check instrument and equipment and submit records.

2 While testing the consistency of channels, the conditions for placement of geophones shall be uniform. The placement area for all geophones shall be very small compared with their distances away from the seismic sources.

3 The insulation of connecting cables and detection cables shall be checked, the insulation resistance shall be greater than 200kΩ.

3.4.5 The layout of survey lines shall meet both the provisions of 3.1.3 in this code and the following requirements:

1 The layout of survey lines shall take into account the side effects and transmission phenomenon.

2 A survey line shall be laid out in a straight line and can be turned while passing buildings, roads, high-voltage electrical lines and other obstacles, but relevant measures shall be taken to ensure that the data for the turned survey line can be independently interpreted.

3 The layout of the survey lines in the valley topography may be sectionalized on the hillside of similar slope. The shot points should be located at the top and bottom of a topographic relief to ensure that the survey line data can be independently interpreted.

4 The survey line at a river valley should be laid out perpendicularly to or along the river. If there are the narrow river valley and short overlapping zones, the survey line can be obliquely crossed with the river.

3.4.6 The test work shall meet both the provisions of 3.1.4 in this code and the following requirements:

1 The information shall be made available in a survey area about geophysical conditions, distribution of effective waves and interference waves, testing measures for suppressing the interference waves, selection of shooting-receiving modes, instrument working parameters and measurement system, etc.

2 The expanding spread measurement technique may be used to understand the distribution of effective waves and interference waves in a survey area. The expanding spread length should be 6 to 10 times the investigation depth for the refraction method, 2 to 3 times the investigation depth for the reflection method, and 1 to 2 times the investigation depth for the Rayleigh wave method. The geophone interval should be less than the actual one.

3 When the poor quality of records from a local section occurs during measuring, a cause shall be found out, so that new instrument working parameters will be selected by testing or an operating practice may be altered to improve the recording quality.

3.4.7 The measuring system shall meet the following requirements:

1 The measuring system shall be determined based on testing results, and under the precondition of fulfilling the requirements for investigation tasks and ensuring the continuous comparison and trace of effective waves, a simplified measuring system shall be used.

2 The requirements for the shallow seismic refraction method are:

1) When using the measuring system for a single forward or reverse shooting, the tracked interface shall be flat and the apparent dip of a layer interface along the survey line direction shall be less than 15° to ensure that there are at least 4 geophone stations that can effectively receive refracted waves.

2) When using the measuring system for forward and reverse shooting, it shall be ensured that there are at least 4 geophone stations that can effectively receive refracted waves in the overlapping zones of a tracked interface.

3) When using the measuring system for overlapping forward and reverse shooting, it shall be ensured that there are at least 4 geophone stations that can repetitively receive refracted waves in the tracked section of the same interface.

4) When using the measuring system for multiple forward and reverse shooting, it shall be ensured that the continuous comparison and trace of refracted waves for each layer can be made, and there are overlapping zones that can be independently interpreted in the composite time-distance curves for stratification.

5) In arrangement of the measuring system for a non-longitudinal survey line, the effect of changes in interface velocity shall be taken into account. The non-longitudinal survey line shall pass through the longitudinal survey line or boreholes and bedrock

outcrops with its length smaller than its distance away from the shot point.

3 The requirements for the shallow seismic reflection method are:

1) The measuring system equipped with one-side or double-side expanding spread may be used to understand the distribution characteristics of effective waves and interference waves within a survey area, and the optimal window of reflections is selected to determine offset and geophone interval.

2) The measuring system for constant offset is applicable to the survey area in which the geophysical conditions are simpler, reflection horizon is more stable and there are stronger internal reflections in the optimal window. During measuring, the center of a reflected wave window shall be selected for the offset based on the testing data of expanding spread.

3) The measuring system for multiple-fold is applicable to the survey area in which the geophysical conditions are more complicated. During measuring, the measuring system shall be used with single-end shot of some offset and no less than 6-fold, which makes the receiving spread within the optimum offset reflections window.

4 The requirements for the Rayleigh wave method are:

1) The stable-state Rayleigh wave method shall have the frequency-conversion vibroseis single-end or double-end used for excitation, the spacing and offset of two geophones adjusted for receiving and records of Rayleigh waves acquired with multiple combinations

of different frequencies.

 2) For the transient Rayleigh wave method, hammer, weight drop and explosion source should be applied, which are excited at the single-end or double-end of a spread, and 12 channels or 24 channels may be used as one spread for receiving, with its offset should not smaller than the geophone interval.

 3) The appropriate offset and geophone spacing shall be selected via testing to fulfill the requirements for the receiving window and investigation depth of the optimal Rayleigh waves. The spread length shall be greater than the investigation depth and the geophone spacing smaller than the dimension of an anomalous body.

 4) The spread direction for the same survey line shall be uniform.

3.4.8 The excitation and reception of seismic waves shall meet the following requirements:

 1 Shooting operation safety shall meet the requirements of GB 6722.

 2 The requirements for explosion seismic sources are:

  1) A clock loop, which is reeled around the explosive charge, shall be used for timing.

  2) The explosive charge shall be buried compacted in a hole. If a shot hole is used for scores of times, loose earth shall be removed every time the explosive charge is buried.

  3) It shall be water-proof for underwater shooting and the sinking depth of the explosive charge shall be about 1m.

4) Water or mud coupling shall be made in borehole shooting.
5) The shot location and depth shall be accurate, and may allow, if necessary, for migrating along a vertical survey line direction, but the migration distance shall not be greater than 1/5 of the geophone spacing.
6) Debris and weed at a shot point shall be cleaned out in surface shooting.
7) Use of 2 or more explosive wires and timing lines are strictly forbidden at a given explosion station.
8) A dedicated blaster is available.

3 The requirements for hammering and weight drops are:
1) The excitation stations for the shallow seismic refraction and shallow seismic reflection shall be located at the more compacted earth, and loose earth at the excitation stations shall be removed or tamped beforehand if necessary.
2) The excitation stations for Rayleigh wave require the ground surface be intact without any tamping or cleaning, so helpful as to excitate low-frequency Rayleigh wave. The strike plate shall be kept in good contact with the ground surface in case the second trigger is generated due to bounce.
3) When a wood board source is used, the long axis of a board shall be normal to a survey line and the center point of the long axis shall be on the survey line or its extension. Sufficient weights shall be added to the board or clasp nails are mounted to hold the board firmly to the ground surface.

4) When a raking source is used, the fixed rake teeth shall be driven into earth, and if necessary, the movable rake teeth can be used to make the rake in good contact with the ground surface.

4 Those methods of shallow seismic refraction, shallow seismic reflection and transient Rayleigh wave usually require 12-channel or 24-channel seismographs to make data acquisition, and the stable-state Rayleigh wave method may require the 2-channel instrument for receiving.

5 The uniform geophone spacing and spread length shall be employed at the same survey line. Geophone spacings should be determined depending on selection of methods, investigation tasks and geophysical conditions.

1) The geophone spacing of the shallow seismic refraction method should be 5m to 10m and accordingly decreased while investigating the weathered zone, fault and rock and soil parameter testing.

2) The geophone spacing of the shallow seismic reflection method should be 2m to 4m.

3) The geophone spacing of the Rayleigh wave method should be 1m to 4m.

6 Inherent frequencies in geophones may be selected according to the technical requirements for effective frequency response and improving resolution.

1) The shallow seismic refraction method should require the vertical geophone with 10Hz to 40Hz inherent frequencies.

2) The shallow P-wave reflection method should require the vertical geophone with 100Hz inherent frequency.

3) The shallow S-wave reflection method should require the horizontal geophone with 40Hz to 60Hz inherent frequency.
4) The geophone inherent frequency and bandwidth of the Rayleigh wave method shall be in conformity with the requirement for investigation depth, and the vertical geophone with the inherent frequencies 1Hz to 10Hz should be used to investigate overlying layers.

7 The requirements for deploying of geophones are:
1) They shall be accurately placed under the uniform buried conditions and with the firm contact with the ground surface, to prevent electrical leakage and background noise.
2) When the geophones are not placed at the original designed points due to the limits of topography and surface conditions, they may be moved along the survey line (not including reciprocal points). If any difficulties, they may be moved along the vertical survey line direction with their offset distances less than 1/5 of the geophone spacing, and recorded in field log.
3) When a horizontal geophone is used to receive S-waves, the geophone shall be horizontally placed, its sensitive axis being normal to the survey line direction and uniformly oriented.
4) When a hydrophone is used in water area for receiving, the hydrophone shall be sunk below the water surface and the depth should be larger than 1m.
5) When a triaxial geophone in borehole is used to receive S-waves, the geophone shall be checked to keep firm

against the wall of borehole before measuring.

8 The requirements for instrument working parameters are:

1) They shall be chosen in accordance with background noise, shooting and receiving conditions, geophysical conditions, safety, etc. in a survey area.

2) When working on the same survey area or measuring section, the same filtering setting shall be used and owing to a special need of changing the filtering setting, the relevant comparison record shall be maintained.

3) The recording length and sampling rate shall be chosen in terms of the characteristics of time domain and frequency domain of effective waves, and in using a high sampling rate for receiving, time delay may be employed as an assistant means.

4) All-pass shall be set in observing transient Rayleigh waves. The sampling rate shall be less than half cycle of the maximum Rayleigh wave frequency, and the time range shall cover the longest travel time of a Rayleigh wave in the farthest channel.

5) When the ratio of signal to noise is low, stacking by repetitive shooting is adopted.

3.4.9 The seismic exploration in water area shall meet the following requirements:

1 When the shallow seismic refraction method is used for observation in the water area, a fixed spread should be employed, with the explosive seismic source and buoyant cable available. In case of torrential water current and strong noise background, the shot station and receiver station can be exchanged.

2 When the shallow seismic reflection method is used in the water area, a moving spread should be employed, using the boat-towed spark source and buoyant cable to move along the survey line in step and keeping the tugboat speed unchanged and the cable sinking depth uniform.

3 When arranging longitudinal survey lines in the cross-river direction and employing forward and reverse shooting observation system, it shall be considered whether the width of riverbed shall cover the overlapping zone as required by the investigation task. The layout of non-longitudinal survey lines in the cross-river direction shall meet the provisions of 3.4.7 in this code.

4 The water level and water depth along a survey line shall be measured timely during operating in the water area. Correction shall be made when the water level variations are over 0.5m.

3.4.10 A certain number of velocity parameters shall be measured in boreholes, adits, outcrops and typical terrains, and the testing methods may include borehole seismic logging, seismic penetrated wave velocity test and elastic wave test on the outcrops and adit wall.

3.4.11 The original records shall meet the following requirements:

1 The original records shall include equipment check record, test data sheet, operational record, field log, etc.

2 Magnetic discs, optical discs and the like for seismic data shall be clearly identified and kept in agreement with the field log book.

3 The shallow seismic refraction method with its expanding spread shall be attached with the complete waveform records and other methods with the waveform records for typical measuring sections as required.

4 The records will be regarded as unqualified when there exists the provision of 3 in 3.1.7 or one of the following defects:

1) Being incapable of reliably tracking records of effective waves.

2) There are abnormal reciprocal or connected channels which would impact the correct comparison and connection of effective waves, and there also are the records that fail to be accurately transferred from other ones at the same shot station.

3) Over 1/6 traces or 2 adjacent traces in same record are abnormal.

4) When it is difficult to identify effective waves or read accurately the travel time because of strong background noise, there are the records of refracted waves, transmitted waves, Rayleigh wave and single-fold reflected waves.

5) There are the records of reflected waves which contain so high background noise that they can not be used to identify the main target layers even after having carried out filtering and horizontal stacking.

6) There are the records which have their record numbers or main content in unconformity with the field log book and difficult to correct.

3.4.12 The comparison of waves shall meet the following requirements:

1 The comparison of waves are primarily designed to identify and track events of effective waves and displacement of waves, which requires choice of the initial phase close to the effective wave and use of single phase or multi-phase contrast that is usually employed in the fissure-developed area.

2 Comparison analysis of effective waves in varied horizons shall be based on such features as similarity of waveforms, proximity of apparent cycles, continuity and coherence of vibrations, attenuation of amplitudes with distance away from the shot point, etc.

3 Displacement of effective waves shall be determined according to the waveform overlapping characteristics while intersection of events for two sets of waves, sharp changes of waveforms and amplitudes and of apparent cycles or apparent velocity, etc.

4 In the observation system for forward and reverse shooting, the contrast analysis of waves for reciprocal and connected channels shall be conducted in terms of the proximity of equalization and dynamic characteristics in travel time of effective waves and the time difference in reciprocal channels of the same spread or connected channels between the spreads shall be less than 3m after a corrected shot depth.

5 The refraction contrast in the observation system with single and multiple reverse and forward shooting shall be based on the parallelism of travel-time curves to identify the refracted waves of the same horizon or varied horizons, and reciprocal time differences in composite time-distance curves shall be less than 5ms via correction and merging.

6 Identification of a reflection event shall be done using the single- or multiple-phase correlation on the records of an expanding spread and the common-source point. The reflected wave and its successive phase or multiple reflection events, which present hyperbola, shall be parallel to each other, and reflection events at varied horizons shall be gradually close up to each other with increase in source-receiver spacing.

7 Contrast analysis of Rayleigh wave shall be carried out in time-domain and frequency-domain, based on their frequency dispersion characteristics. The Rayleigh wave events reflecting variations of horizons shall be gradually fallen away with increase in source-receiver spacing, and time differences between events are steadily increased with reduced frequencies.

3.4.13 Readings of travel times shall meet the following requirements:

1 The reading of the first arrival time of waves shall be taken from the original records. When it is difficult to take a direct reading of the first arrival, the time for the first extreme reading of effective waves may be taken, but a phase correction is necessary.

2 Correct readings shall be taken while interference and displacement of waves exist through analysis of wave superposition features.

3.4.14 Readings of travel times shall be corrected in terms of Equation (C.4.2) of Annex C in this code and its items should include phase correction, shot depth correction, near-surface low velocity zone and topographic correction.

3.4.15 Readings of average velocities and effective velocities shall meet the following requirements:

1 Effects of relative changes in near-surface inhomogeneity of media, low velocity zone and underlying layer thickness shall be taken into account in determining average velocities and effective velocities.

2 Velocity parameters may be obtained from the measurements of borehole seismic logging, shallow seismic reflections, shallow seismic refractions and Rayleigh waves.

3 Average velocities shall be calculated utilizing borehole seismic logs in the survey area and effective velocities obtained by the methods of square co-ordinates. Crossing points can be assessed and corrected.

4 For seismic exploration, tests of effective velocities should be performed per 100m at two ends of every measuring section. When the adjacent velocities difference is found greater than 20%, additional tests of velocities in this measuring section shall be done, and a curve of velocity variations along a survey line can be plotted, on which corresponding velocity values are used to construct an interface.

5 The low velocity zone correction shall be first done when variations in low velocity zone thickness cause significant variations of effective velocities, and then an interface can be constructed using the effective velocities of underlying layers in the low velocity zone.

6 On the same survey line, if a sharp change in effective velocity is not justified by using sufficient data, no interface shall be constructed using effective velocities in section to avoid sharp change in interface depth.

3.4.16 In addition to the provisions of 3.1.8 and 3.1.9 in this code, data processing and interpretation and plots shall also meet the following requirements:

1 Scales for mapping time-distance curves shall be chosen according to actual accuracy of observation. Horizontal scales for man-made mapping are preferably 1 : 1000 or 1 : 2000 and 1cm in vertical scales stands for 10ms or 20ms.

2 After a time-distance curve has been mapped, it may be checked in terms of the fundamental of equalization of reciprocal time, parallelism of overlapping forward and reverse T-D curves and equalization of intercept time on both sides of a shot point. When there is an anomalous phenomenon, readings of seismic records for relevant channels shall be checked and corrected.

3 When a sharp change in travel time occurs in an individual channel, it shall be checked by comparing to the travel time of single or overlapping forward and reverse T-D curves in relevant terrains, so that a necessary correction may be done as the cause is found out.

4 A composite time-distance curve shall be plotted along with the observed time-distance curve after correction of shot depths and phases.

5 Interface velocities and depths shall be obtained by interpretation of forward and reverse T-D curves. Interface depths can be obtained using the techniques of intercept time for single forward or reverse time-distance curves or critical distance only when there are approximately horizontal layered media, small undulating ground surface and interfaces, and no significant variations in lateral velocities.

6 The techniques for constructing shallow seismic refraction interfaces shall be chosen according to geophysical conditions and the peculiarities of interpretation methods and accuracy requirement. Interpretation of a single forward or reverse time-distance curve is the intercept time method, critical distance method, and forward fitting computation techniques. Interpretation of forward and reversed curves is the t_0 method, delay time method, time field method, conjugate point method and forward fitting computation techniques. The requirements for choice of the techniques are:

1) The t_0 method or delay time techniques may be

employed when there are less undulating interfaces, no penetration phenomena and no significant variations in interface velocities along a survey line.

2) The time field technique may be employed when there are a given undulating interfaces, larger undulating refraction interfaces, no penetration phenomena and significant variations in interface velocities.

3) The conjugate point technique may be employed when there are flat ground surface, larger undulating refraction interfaces, no penetration phenomena and no significant variations in interface velocities.

4) For the inhomogeneous multi-layers with special structures, variety techniques may be used to construct an interface or stratification.

7 Geologic interpretation shall be made based on analysis of data associated with relevant geology, drilling and other geophysical exploration as well as task requirements:

1) The data associated with boreholes or physical properties shall be used to determine the corresponding relationship between the seismic interface and geological interface.

2) Lithological changes in the horizontal direction shall be inferred based on physical property and geological data.

3) The corresponding relationship between the low velocity zone and fault fractured zone shall be established through analysis of whether or not there are accompanying amplitude attenuation and waveform variations on original records.

8 The requirements for result plots of the shallow seismic

refraction method are:

    1) The result plots should include composite time-distance curves map, geologic interpretation result profile map or plane map.

    2) At the upper part of the composite time-distance curves map, the composite time-distance curves and observed time-distance curves shall be plotted also with a complete interpretation auxiliary line. Furthermore, the intercept time, surface layer velocity and effective velocity shall be indicated above the shot point, and at the below part, the seismic geologic profile shall be plotted with the geological interface and fault structure, the interface velocity also be indicated.

    3) The geologic interpretation result profile shall be annotated with scale, elevation, stake number and orientation of a section, end points and turning points of a section, crossing points of a survey line, wave velocity and geologically lithological symbols of upper and lower media at an interface, on which are marked the locations of survey points passing through the survey line. Scales shall conform to the requirements for measuring accuracy of seismic exploration.

    4) The geologic interpretation result maps may include isopach maps of overburden, contour maps of bedrock surface, and isopach maps and distribution maps of interface velocity for other layers of interest. On the maps shall be shown the geologic boundary and interpretation structure line, survey

> line and serial number, location of borehole and number of hole, location and number of pits, signs of main topographic features, etc.

3.4.17 In addition to the provisions of 3.1.8 and 3.1.9 in this code, data processing and interpretation and plots for the shallow reflected wave method shall also meet the following requirements:

1 The forward and inversion coherent analysis shall be made employing data of expanding spreads and known geological and geophysical information to determine reflected wave trains of layers of interest.

2 The processing flow shall be worked out in accordance with the ratio of signal to noise ratio on original records and investigation tasks, followed by choice of such processing parameters as filtering frequency, filtering apparent velocity, stacking velocity, average velocity, etc.

3 Pre-processing of original records includes data zeroizing of abnormal channels and correction of reversed polarity channels.

4 In the survey area where there are large topographic relief and great variations in velocities or thicknesses of near-surface low velocity zones, the static correction for one time shall meet the following requirements:

> 1) When there are large topographic reliefs and apparently different topographic units, base lines shall be corrected in terms of topographic units and by sectionalized choice of topography.
>
> 2) Small additional refractions as required shall be observed along a survey line, at the direction of which there is a larger change in near-surface low velocity

zones.

5 When analysis of velocities is carried out, the requirements for choice of stacking velocities are:

    1) Stacking velocities may be obtained using velocity spectrum or velocity scanning and if any complex geophysical conditions, those two techniques are available for mutual check.

    2) Velocity scanning shall be chosen using less velocity increment in the relatively flat topography with the high ratio of signal to noise in seismic records.

    3) The measuring sections along a survey line shall be sufficient to carry out the analysis of velocities, plot a velocity expansion map and to study the lateral change regularity of velocities by combining velocity logs.

    4) For the terrain where horizontal stacking effect is not good enough, the necessary correction of its stacking velocities shall be done.

6 The requirements for digital filtering are:

    1) Choice of filtering frequencies shall be made on the basis of spectral analysis.

    2) Width of filtering shall be adjusted so as to settle the discrepancy between the enhanced ratio of signal to noise and the resolution.

    3) Proper edge width (offset distance) shall be chosen to avoid Gibbs phenomenon and decrease the error caused by filtering factor truncation.

    4) In the structurally-developed area, pre-stacking two-dimensional filtering is not preferred to avoid the effects of horizontal mixed waves on seismic

dynamical peculiarities at fault.

7 Based on the geophysical conditions and seismic record features in a surveyed area, the processing techniques shall be chosen as follows:

1) The dynamic equalization processing of the records that present strong amplitude contrast between effective reflected wave trains, shall be performed prior to stacking.
2) The deconvolution should be made of the records of mutual interferences between effective reflected wave trains.
3) The corresponding NMO (normal moveout) correction should be made of the measuring section with vertical velocity reversal.
4) When the interface dip is large, offset stacking or stacking offset can be done.

8 Profile finishing processing should not be preferred in the vicinity of fault-developed zones and fractured zones.

9 The NMO correction shall be made of original constant offset data or seismic image data for quantitative interpretation.

10 The requirements for basic data interpretation plots of shallow seismic reflection are:

1) Such items as the name of survey area, number of survey line, offset, and geophone spacing shall be indicated on plots.
2) Such items as the stacking fold, processing flow, stacking velocity, etc. shall be indicated on horizontal stacking time profiles.
3) Whether the NMO processing has been carried out and the NMO velocity values has been marked shall be

indicated on equal offset time profiles or seismic images.

4) The time profile for a typical terrain shall be attached with relevant expanding spread records.

11 The contrast analysis shall be done employing geological and other geophysical exploration data in terms of basic plots to determine the relationship between geological horizons and seismic wave trains. The wave trains in response to an investigated layer of interest can be chosen for the contrast and trace to obtain characteristics of reflectors. The number of reflectors shall be annotated.

12 The interpretation of a variety of time profiles shall include identification of the relationship between principal geologic horizon and reflector and of variations in layer thickness and contact relation, demarcation of faults or fractured zones.

13 The analysis of such phenomena as bifurcation, merging, discontinuity and pinch-out of wave trains in time profile shall be made to find out the relationship the phenomena and the variation of layers, lithologies and structures.

14 Sedimentary structures and other geologic evidence in the Quaternary unconsolidated strata shall be identified and interpreted with geological or other geophysical data for comparison and proof.

15 The relationship shall be analyzed between the waveform dynamical characteristics in the constant offset time profile or seismic images and the variations of subsurface media in horizontal and longitudinal directions. The dynamical characteristics include amplitude, frequency, phase, after-vibration, wave dispersion of waveforms.

16 The techniques for constructing shallow seismic reflection interfaces, which involve cross point, ellipse, circle, time field, etc. shall be chosen according to geophysical conditions and the features of and accuracy requirement for interpretation methods.

  1) When the overburden are homogenous and the variation in their average velocities are not large, the cross point, ellipse and circle techniques can be used to construct a reflection interface. The cross point and circle techniques are applicable to horizontal and inclined interfaces and the circle technique applicable to the interface less than 10° in apparent dip.

  2) When the overburdens are inhomogeneous and their average velocities change significantly, the time field technique can be used.

  17 The result plots include geologic interpretation result profiles or maps, depth contour or equivalent t_0 maps, and distribution maps for fault structure lines.

3.4.18 In addition to the provisions of 3.1.8 and 3.1.9 in this code, data processing and interpretation and plots for the Rayleigh wave method shall also meet the following requirements:

  1 The time or phase difference techniques may be chosen to calculate V_R in a layer within $\lambda/2$ depth range below the surveyed surface for a given frequency of the steady-state Rayleigh wave.

  2 Stable state Rayleigh wave velocities shall be calculated by choice of the records on the same side of a shot point and at these two geophone stations with the phase difference in $2\pi/3$ to 2π, taking readings one by one of time difference or phase difference-calculated V_R of Rayleigh waves, from high frequency to low frequency, and a dispersion curve is mapped at the test point which is located at the center of the connecting line of those two geophone stations.

3 The transient Rayleigh data processing flow shall include the four steps of time-distance, wave number of frequency, distance frequency and depth velocity, and further, reduction, refinement, stacking and inversion, of data, and presentation of intermediate and final processed results in image and data sheet.

4 In data processing of transient Rayleigh wave, first analyze the peculiarities and distribution of Rayleigh wave frequency dispersion, choose the time window for spectral analysis of amplitude spectrum and phase spectrum, extract the Rayleigh wave of different frequencies from various seismic channels within the time window and use sound processing methods to derive Rayleigh wave dispersion curves.

5 Such techniques as cross correlation, phase difference, frequency-wavenumber domain and spatial autocorrelation, shall be used to calculate Rayleigh wave velocities. Such techniques as the extreme-value or approximation-point and first-order derivative or inflection-point can be used to obtain layer thickness. S-wave velocities can be calculated according to Equation (C.4.5) of Annex C in this code.

6 The traits of "inflection-points" of dispersion curves for Rayleigh wave data can be used to interpret the interfaces in subsurface elastic media and for the profile survey, magnitude and distribution of a velocity anomaly area for Rayleigh waves in a profile plot can be employed for interpretation.

7 Half wavelength technique may be used for depth conversion of Rayleigh wave, which may have coefficients of Poisson ratio corrected and contrast interpretation made in reference to geological data in a survey area.

8 For the dispersion curve, a wave velocity frequency curve shall be mapped with the Rayleigh wave frequency as the longitudinal axis and the Rayleigh wave Velocity as the transverse axis, and also a depth frequency curve can be done as such. A map can be plotted with short broken line for the steady-state method, with dot curve for the transient method and also the same map plotted at one time with short broken line.

9 When carrying out profile testing, wave velocity frequency or depth frequency curve is to be plotted on the same profile at a scale as required by a task, contour map of V_R profile, chromatogram map and grey-scale image can be plotted in terms of velocity and depth of a layer for the final inversion calculation.

3.5 Measurement of Elastic Waves

3.5.1 Measurement of elastic waves may use the acoustic and seismic wave techniques, the former including the single-hole acoustic wave, penetrating acoustic wave, surface acoustic wave, acoustic wave reflection, pulse echo and the latter including the borehole seismic logging, transmitting seismic velocity testing and continuous seismic velocity testing.

3.5.2 The application conditions shall meet the following requirements:

1 The single-hole acoustic waves shall be measured in the borehole with well fluid coupling and without metal casing.

2 In the cross-hole measurements of penetrating acoustic waves, there should be well fluid coupling with a hole distance so appropriate as to ensure that the receiving signals are clear.

3 The testing of surface acoustic waves, acoustic wave reflections and continuous seismic wave velocities shall be conducted on the smoother surface of concrete, bedrock outcrops, exploratory trenches, shafts and cavities.

4 The pulse echo shall be applied to the wave impedance face significantly existing in an object and its surrounding media, on which multiple echo signals can be generated inside the object.

5 The borehole seismic logging should be performed in the borehole without metal casing.

6 The testing of penetrating seismic velocity should be conducted in the borehole, adit or open face. For the purpose of investigating, there shall be a significant difference in the wave velocity between the surveyed object and its surrounding media with a given magnitude.

3.5.3 The instruments shall meet the following requirements:

   1 The requirements for acoustic apparatus are:

      1) The minimal sampling interval is $0.1\mu s$.

      2) The sampling length is not less than 512 sampling points per channel, which is optional.

      3) The triggering modes should be the internal, external, signal, stable-state, etc.

      4) Band width is 10Hz to 500kHz.

      5) The measuring accuracy of acoustic time is $\pm 0.1\mu s$.

      6) Transmitting voltage is 100V to 1000V.

      7) The transmitting pulse width is $1\mu s$ to $500\mu s$, which is optional.

   2 The floating-point amplifier with wave train presentations should be used in the acoustic wave reflection, including controllable and better uniform source energy, good frequency characteristics for a receiving transducer and moderate damping.

 3 The instruments, which have the wide frequency band, high sampling rate, great sampling length and spectral analysis function, shall be used in the pulse echo.

 4 The requirements for using electric sparks and supermagnetostriction seismic sources are:

  1) The protection and use of instruments shall conform to the requirements for high-voltage electrical apparatus.

  2) The seismic source shall excitate high-frequency acoustic wave pulses and control energy.

  3) The clock signals skip sharp and stable, synchronizing with the receiving instrument and the time-delay error shall be 2 times less than the reading error.

 5 The instruments for the seismic wave method shall meet the provisions of 3.4.3 in this code.

3.5.4 The preparation of fieldwork shall meet the following requirements:

 1 Before testing acoustic waves, check acoustic instruments that include the triggering sensitivity, probe performance, cable marker, etc.

 2 The cylinder transmitter and receiver probes in a pool shall have different spacings for measurements. The zero value can be taken by plotting the curves for 3 to 4 stations and directly measured by a plane probe using a coupling agent.

 3 The testing of seismic waves shall meet the relevant provisions of 3.4.4 in this code.

3.5.5 The operational arrangement shall meet both the provisions of 3.1.3 in this code and the following requirements:

 1 In measuring bedrock outcrops, exploratory trenches,

shafts and caverns, the measuring section shall be laid out on the representative and flat terrain in accordance with the requirements of investigating tasks. The successive profiles in the underground caverns of magmatic rock and thick rock mass shall be measured in layout of survey lines at the same height of the cavern walls. For stratified rock mass, the survey lines may be laid out along the same layer. The station spacing shall be determined based on geophysical conditions and the requirement for accuracy and resolution of instrument clock, and it shall be 0.2m to 0.5m for the acoustic method and 1m to 2m for the seismic wave method.

2 In case of surface acoustic waves, acoustic wave reflections and pulse echoes, suitable survey grids and working scales shall be chosen to ensure that the minimal anomalies as required by measuring tasks can be detected, clearly reflecting the position and shape of a surveyed object.

3 In measuring penetrating acoustic waves or penetrating seismic velocities, the hole distance shall be defined in terms of geophysical conditions, instrumental resolution and excitation energy, including measurement of the borehole deviation and correction of the hole distance.

4 In conducting the comparison test on dynamic elastic modulus and static elastic modulus, the boreholes shall be laid out along the different directions of a test-piece plane of static elastic modulus, including the proper arrangement of the number, distances and depths of the holes and measuring the acoustic velocities of a rock mass along the different directions.

3.5.6 The onsite measurement shall meet the following requirements:

1 The requirements for in-hole measurement are:

1) First check the accessibility of the borehole using a probe-like object which is slightly larger than the test probe in diameter and weight. In case of great inclination or up inclination borehole, it is better to use a probe-like stick.
2) Cable depth mark shall be accurate and conspicuous.
3) When there is casing in a borehole, the annulus outside the casing should be infilled with water, sandy soil and so on.

2 The requirements for measurement of single-hole acoustic waves are:

1) The acoustic probe with one transmitter and two receivers should be utilized.
2) In carrying out measurements of acoustic waves in a dry hole, a dry-hole acoustic probe shall be used and kept in good contact with the wall of hole and maintain distinct receiving signals.
3) Measurements should be taken from the bottom to the top of a hole, their spacing is 0.2m and a depth correction is made every 10 points.
4) High-power transmitter probes or receiver probes with pre-amplification should be used for the relatively fractured or deeper borehole.

3 The requirements for measurement of penetrating acoustic waves are:

1) The measuring mode may be horizontal synchronizing, inclined synchronizing, etc.
2) When measuring the concrete mass or rock mass that has two relative open faces, a plane acoustic wave probe with proper frequency shall be chosen and

coupled with a couplant. The distance between the transmitter and receiver shall be accurate determined.

4 The requirements for measurement of surface acoustic waves are:

1) The plane acoustic probe shall be used.
2) Single or forward and reverse shooting measuring system may be used to measure acoustic velocities.
3) High-power transmitter probes or receiver probes with pre-amplification may be used when the distance is larger or acoustic attenuation is faster.
4) The location of a probe shall be flat, coupled with a couplant.

5 The requirements for measuring acoustic reflections and pulse echoes are:

1) A test shall be done in the known terrain, through choice of suitable offset, excitation energy, instrument parameters, etc.
2) A plane acoustic probe shall be used to measure constant offsets and the acoustic wave receiver probe shall be characterized by high sensitivity and intermediate damping.
3) For acoustic reflections, external triggering sources with narrow pulses may be chosen such as magnetostriction, rebound hammer, etc. and for the pulse echo, the rebound balls with different frequencies may be chosen according to the measuring requirements.
4) The surface for placement of a probe shall be flat, coupled with a couplant.

6 The requirements for measuring continuous seismic wave velocities are:

 1) The less undulating terrain for rock mass should be chosen, on which the survey lines shall be laid out in terms of geological structures, lithology, weathering and integrity of rock mass.

 2) The spread length shall be determined according to the integrity of a rock mass and instrumental reading accuracy. The P-wave travel time for the adjacent channels in the spread shall surely be 5 times larger than the instrumental reading accuracy.

 3) Single or forward and reverse shooting measuring system should be used.

 4) In measuring the s-wave velocity of a rock mass, a wood board should be used at the bedrock outcrop. A wall support should be used in the cavern with positive and reverse excitation, and it is required that the acquisition rate for S-waves be no less than 60% of P-waves.

7 The requirements for borehole seismic logging and measuring seismic wave transmission velocity are:

 1) The shooting and receiving shall meet the relevant provisions of 3.4.8 in this code.

 2) For borehole seismic logging, the shot at the ground near the borehole and receiver inside the borehole measuring system may be chosen.

 3) The distance between the surface shot point and the hole top can be determined by a test and should be 2m to 4m.

 4) A station spacing shall be determined by the wave velocity of a layer, which is 1m to 2m for overburden

and 2m to 3m for bedrocks.

5) The simultaneous observation should be used to measure cross-hole penetrating seismic velocities along the strata dip direction. When taking measurements in three holes at the same straight line, what may be used includes one hole for shooting and two other holes for receiving, and the distance between the shot point and the receiver station should be corrected in terms of borehole inclination survey data.

6) An in-hole shear hammer source should be moved from bottom to top and pushed firmly against the wall with its depth being accurate.

7) When a three-component geophone is used for measuring S-waves, it shall be adhered to the wall.

8) A spark source should be used for measuring cross-hole penetrating seismic velocities, coupled with water or slurry, and when the hole distance is larger, an explosive seismic source may also be chosen.

9) For the measurement of cross-adit or cross-open face penetrating seismic velocities, an explosive seismic source or a hammer source may be chosen, depending on distances or geophysical conditions.

3.5.7 Measurement and repeated observation shall meet the following requirements:

1 Appropriate attenuation or gain settings shall be chosen while reading, so as to make amplitudes proper and the first arrival point or reflected waves distinct and readable. In measuring amplitudes, keep the measuring conditions unvaried and take readings of the amplitude values at the same phase, indicating the phase in reading.

2 When intermittent shooting or hammering is adopted, the distinct and stable first arrival and reflected signals shall be chosen through analysis of their waveforms.

3 The repeated observations of the measured section, in which there are sharp changes in the wave curve or skips at the station, shall be taken through overlapping or increasing shot energy, with the mean value of 3 repeated observations as the measured result.

3.5.8 The data processing and interpretation shall meet both the provisions of 3.1.8 in this code and the following requirements:

1 Before analysis of interpretation of results, make corrections of the zero point, hole deviation, height difference, offset, etc.

2 The coefficient of rock mass integrity shall be calculated using the same wave velocity of a fresh, intact rock mass for the same lithology in a survey area, which can be calculated as per Equation (C.6.3) of Annex C in this code and assessed according to the requirement of Table 3.5.8.

Table 3.5.8 The classification of coefficients of rock mass integrity

Level of integrity	Integrity	Less integrity	Poor integrity	Less shattered	Shattered
Coefficient of rock mass integrity	$K_V > 0.75$	$0.75 \geqslant K_V > 0.55$	$0.55 \geqslant K_V > 0.35$	$0.35 \geqslant K_V > 0.15$	$K_V \leqslant 0.15$

3 The dynamic elastic modulus may be calculated using the acquired longitudinal wave velocity, S-wave velocity and density value and its relationship may be established via dynamic-static comparison.

4 The spectral analysis of pulse echoes shall be conducted by comparison analysis of various peak values and finding out the layer of interest or defective echo frequency, whose thickness or defective buried depth may be calculated based on measured velocity parameters.

   5 The analysis of a wave train for a penetrating acoustic wave shall be performed.

3.5.9 The result plots shall meet both the provisions of 3.1.9 in this code and the following requirements:

   1 The integrated measured result plot shall be mapped using cavern elastic wave testing, which includes the wave velocity curve, dynamic elastic modulus curve and integrity coefficient curve.

   2 Wave velocity curves together with statistical analysis curves shall be all mapped for single-hole acoustic waves, penetrating acoustic waves and borehole seismic logging while conducting engineering quality detection. When multiple holes (cavities) are located in the same profile or cross-section, the wave velocity curves shall be mapped in the same profile or cross-section.

   3 When a survey grid consists of the multiple survey lines used to measure the velocities of surface acoustic waves and continuous seismic waves, a distribution map for wave velocity shall be plotted.

   4 The constant offset-time profiles and the geophysical exploration interpretation profiles or plans shall be plotted for acoustic reflections.

   5 The time and frequency profiles shall be mapped for pulse echo.

3.6 Computerized Tomography

3.6.1 The computerized tomography (CT) involves elastic

wave velocities and electromagnet wave absorption coefficients (electromagnetic wave CT for short), and the elastic wave velocity CT includes acoustic wave velocity CT (acoustic wave CT for short) and seismic wave velocity CT (seismic wave CT for short).

3.6.2 The application conditions shall meet the following requirements:

1 There is a difference in electrical property or elastic wave velocity between the surveyed object and its surrounding media. The electromagnetic wave CT shall be used for the difference in electrical property, while the acoustic wave CT or seismic wave CT shall be used for the difference in elastic wave velocity. Provided the differences in electrical property and elastic wave velocity exist simultaneously, one of the above-mentioned CT methods can be chosen based on the conditions, and multiple CT methods can be used in case of complicated conditions.

2 The investigation conditions for boreholes, adit and open faces shall be available in the vicinity of or at least on both sides of an imaging area, and the surveyed object shall be relatively located at the center of a scan profile, and its dimension shall have comparability with its imaging element.

3 High-frequency acoustic waves are suited to cross-section scanning of relatively small area and low-frequency seismic waves to cross-section scanning of relatively large area.

3.6.3 The instruments shall meet the following requirements:

1 The instruments for acoustic wave CT shall meet the relevant provisions of 3.5.3 in this code.

2 The instruments for seismic wave CT shall meet the relevant provisions of 3.4.3 in this code.

3 The instrument for electromagnetic wave CT requires

that its frequency be optional, its noise level be less than 0.2μV, its measuring range be 20dB to 140dB, its dynamic range be 100dB and its measuring error be less than ±3dB.

3.6.4 The layout of stations shall meet the following requirements:

1 The CT profile shall be normal to the trend of a layer or surveyed object and the borehole and adit in a scan profile shall be coplanar and relatively regular.

2 The holes (cavities) shall be duly spaced in terms of the requirement for investigation tasks, physical property conditions, instrumental performance and features for investigation techniques, with its spacing less than 30m for acoustic wave CT, less than 40m for electromagnetic wave CT and properly selected for seismic wave CT based on the excitation mode and energy magnitude. The depth of a hole (cavity) that is imaged shall be larger than its spacing, and in the section where the geophysical conditions are relatively complex and the higher survey accuracy is required, the hole or cavity spacing shall be correspondingly decreased.

3 The station spacing shall be chosen according to the requirements for survey accuracy and techniques, which is not greater than 1m for the acoustic wave CT, 1m for the electromagnetic wave CT and not greater than 3m for the seismic wave CT.

3.6.5 The measuring system shall meet the following requirements:

1 Two-sided observation systems can be used for CT in between holes (cavities), three-sided observation systems can be used when the conditions for a cross-hole ground surface or a side slope are appropriate, and many-sided observation systems can be used when there is a beam column or free-face body.

2 Covering adits, boreholes and natural open faces, a seismic wave CT profile shall be constructed by making full use of the excitating and receiving conditions around a survey area and adopting a fan-shaped observation system with one transmitter and multiple-receivers to keep uniform distribution of rays and crossing angles not much less.

3 A fixed-point fan scan should be mainly used for observations of acoustic wave CT and electromagnetic wave CT, supplemented by horizontal synchronizing and tilted synchronizing observations. The maximum angle for fixed-point scanning observation depends on the principle that there is not a significant diffraction generated beyond the profile.

4 When the spacing of the transmitter station is larger than that of the receiver station, an observation method is available with exchange of two holes and at least of a given number of transmitter stations and receiver stations.

5 When performing multiple cross-hole or cross-cavity CT observations, the observation system should be kept uniform.

3.6.6 The fieldwork shall meet the following requirements:

1 When performing a CT observation in a borehole, a complete understanding of the borehole shall be made and relevant precaution be taken to reduce occurrences of hole accidents.

2 The acoustic wave CT and seismic wave CT shall meet both the provisions of 3.5.6 in this code and the following requirements:

　　1) The acoustic log and inclination survey shall be carried out in the borehole available for the acoustic wave CT or seismic wave CT, and the measurements of seismic waves or acoustic velocities shall be made in the adits

available for the seismic wave CT.

2) The seismic wave CT or acoustic wave CT in a borehole requires the hole with relatively unbroken wall be chosen as the receiver hole, and when the conditions for hole wall are poor, plastic casing may be run.

3) When an explosive seismic source is used to carry out the cross-hole seismic wave CT, the metal casing shall be run into the shot hole for wall protection, and the shots below the casing shoe are being fired while lifting the casing from bottom to top, so that collapse of the hole wall would be prevented.

4) When the distance between the shot and the receiver is a little far, the high-energy excitation apparatus or pre-amplification receiver probe or high-sensitivity geophone shall be used for receiving.

3 The requirements for electromagnetic wave CT are:

1) The instrumental working frequency and corresponding antenna shall be chosen through a field test.

2) The optional observation modes include single frequency and multi-frequency. The same frequency shall be used for multiple sets of electromagnetic wave CT in one cross-section, and in the hole interval under relatively simple geophysical conditions, a three-hole method can be used to carry out simultaneous observation to determine the initial field intensity and background value.

3) An insulated cord, which has a length of 2 times the chosen wavelength, should be used to connect the transmitter. A filter shall be used to connect the

receiver. A weight shall be attached to the antenna respectively.

4) When electromagnetic wave CT is carried out in between holes (cavities), effects of such metal pieces as steel pipes and rails shall be avoided, and a metal plate will be used to shield the holes (cavities) to prevent the diffraction of electromagnetic waves.

3.6.7 The repeated observations of sharp anomalous changes shall be done, a check observation shall be performed of the stations in an anomalous area, the observational data for exchanging the transmitter and receiver can be taken as check workload.

3.6.8 Relative error δ for the single-point repeated observations of travel time or field intensity shall be less than 3.5% and relative mean square error m for the check observations shall be less than 5%.

3.6.9 In addition to the provisions of 3.1.8 in this code, the data processing and interpretation shall meet the following requirements:

1 A coordinate system shall be established according to survey data, and excitation points and receiving points for each ray would be converted into imaging profiles of a two-dimensional co-ordinate, which, combined with the relevant travel time or field intensity data, are incorporated into a data file.

2 The average velocity or mean absorption coefficient for each ray shall be calculated, all the relevant synchronizing and fixed-point parameter curves be separately displayed, and the position of an anomaly and the range of variations in inversion parameters be basically identified.

3 A mathematic physical model shall be established based on geophysical conditions, observation systems, imaging accuracy, resolution and investigation tasks to determine the shape and size of elements and nodes. The size of an element grid shall be greater than the station spacing and the total number of elements can be less than the rays.

4 A ray should be traced from a straight line to a bent line, using the linear equation, target shooting, minimal travel time or square slowness, etc.

5 The inversion algorithm may use singular value decomposition (SVD), simultaneous iterative reconstruction technique (SIRT), conjugate gradient (CG), damped least square method (LSQR) and other improved above-mentioned techniques.

6 The number of inversion iterations shall be defined in terms of the stability of ray paths and image shapes and also in terms of the image data variance of two adjacent iterations.

7 Related data from in-situ measurements in boreholes or adits, as boundary conditions, may be put into relevant inversion calculations to remove the artifacts result from boundary effects.

8 For the interconnected CT profiles, the same inversion technique, model and parameters shall be used.

9 The final ray distribution map for bent line inversion can be used as one of results. The positions and dimensions of high-velocity or low-velocity zones and high absorption or low absorption zones can be determined according to ray density, and the area and elongating orientation of anomalies can be determined in terms of varying gradients of CT image parameters.

10 The inferred geological interpretation shall be done based on CT image velocities or distribution of absorption coefficients and combined with layer lithologies and geological structures in the survey area.

3.6.10 The result plots shall meet both the provisions of 3.1.9 in this code and the following requirements:

1 The plots should include CT image, ray distribution map, CT interpretation result plot, and the result plots mapped from other related measured data.

2 Such graphic techniques as contour, shade of grey, color spectrum, etc. may be employed in CT images that are graded in equi-difference and also with variation.

3 Multiple sets of CT profiles in the same cross-section may be merged into a result profile.

3.7 Sonic Echo Exploration

3.7.1 The sonic echo exploration may be used to investigate the underwater topography in reservoirs, river courses, lakes and the deep water areas of shallow sea for water resources and hydropower project, as well as the underwater stratigraphic cross-sections in the projects associated with dam sites, bridge foundation and ports.

3.7.2 The application conditions shall meet the following requirements:

1 There is a wave impedance difference between the surveyed layer and its adjacent strata.

2 In case of stratification of overburden below water bottom, the target layers shall be characterized by some thickness, uniform media and steady wave velocities.

3 There is no gravels or a sparse distribution of them

above the surveyed layer of interest.

3.7.3 The instruments shall meet the following requirements:
    1 Investigation depth and resolution:
        1) When the water depth is less than 50m, and the loose sedimentary layer thickness is about 0m to 25m, the resolution is 0.3m to 0.5m.
        2) When the water depth is greater than 50m, and the loose sedimentary layer thickness is about 25m to 50m, the resolution is 0.5m to 1.0m.
    2 Oscillation number:
        1) Shallow: 0m to 25m, 360 times/min.
        2) Deep: 0m to 50m, 180 times/min.

3 The gain of a receiver amplifier shall be not less than 150dB.

4 The sensitivity of a receiver probe shall be not less than $10\mu V/\mu bar$.

3.7.4 The survey line layout shall meet the following requirements:

1 The survey lines for river courses and reservoirs shall be normal to the trend of underwater topography and the layout of a cross-river profile is preferable with its line spacing being 50m.

2 When the underwater topography is a little flat, the survey line may be laid out along the down current direction.

3.7.5 The measure of location shall meet the following requirements:

1 The observation profile line requires the use of GPS positioning measurements and of other positioning measuring methods for the works in river courses or reservoirs.

2 GPS measuring method shall serve as a real-time dynamic measurement.

3 For the instruments for analog signals, hydro-acoustic record marking and GPS data sampling shall be simultaneously done, and the spacing between positioning points can be defined as per the survey accuracy, and shall be less than 50m. For the instruments for digital signals, GPS successive positioning can be employed.

4 In a profile survey, the error of measurements shall be less than 2mm at the co-ordinate of an available topographic map, the elevation less than 0.5m, and the relative error of the profile distance between both banks less than 1%.

3.7.6 The fieldwork shall meet the following requirements:

1 The field work shall use the flat-bottom motorboat that has the deadweight tonnage of 10t to 20t and is small in noise. In order to ensure smooth sailing of the motorboat, the minimal operating depth shall be determined according to the draft depths, which require that the water depth be preferably greater than 2m.

2 The transmitter probe and receiver probe shall be separately mounted on both sides of a motorboat, the former at the rear of one side and the latter at the front of the other side, the distance between which shall be 6m to 8m, and the immerged depth for the probes is reliable on the magnitude of water wave and shall be 0.5m at the time of undisturbed water surface.

3 For the tow fish probe with the transmitter and receiver all-in-one, it shall be horizontally mounted and a little away from the motorboat to prevent it from bumping the hull and interference on underwater signals.

4 The surveying motorboat shall steer along a profile in a fixed direction and at a constant speed, and the instrument operator shall write marks in the required time.

5 In the process of observation, appropriate frequency parameters shall be chosen according to variety strata and by means of an experiment to enhance stratigraphic resolution. The receiver's sensitivity and conversion frequency shall be adjusted with changes in the depth.

6 A measurement of water level shall be done each day before starting a hydro-acoustic survey in river courses and water reservoirs. When fluctuations in water level are great (the elevation difference is greater than 0.3m), water level shall be measured at regular intervals, with their curve plotted with time during the work.

3.7.7 Data processing and interpretation shall meet the following requirements:

1 When water depth is less than 15m and a split probe is used, course corrections shall be made to eliminate the depth errors caused by the migration of transmitters and receivers.

2 The calibration shall be made of the velocities of underwater surveyed layers to accurately calculate the thickness of each surveyed layer.

3.7.8 The result plots shall include underwater topographic maps, sonic echo geological profiles, isopach map for silt seam or overburden, contour maps for bedrocks, etc.

3.8 Radioactivity Survey

3.8.1 Radioactivity survey includes natural gamma (gamma-ray for short), alpha-ray, measurement of environment radon concentration, isotopic tracer method, etc.

3.8.2 The application conditions shall meet the following requirements:

1 Gamma-ray method is not restricted by topography, but

the geometric configuration shall be uniform in measuring, which shall be used in the following geological conditions:

> 1) There is a significant radioactivity contrast between the surveyed object and its surrounding layers.
> 2) Those places such as fractured zone and shallow subsurface water storage structures.
> 3) Quaternary overburden which have no "shield layers" such as phreatic stratum.
> 4) Magmatite area.

2 Alpha-ray method can be conducted on the overburden soil sample or buried static electric α-card, and has the same conditions as gamma-ray. But it is inappropriate to use the alpha-ray method in overcast and rainy seasons.

3 Collection of samples in measuring radon concentration in air shall conform to the provisions of GB/T 14582.

4 The isotopic tracer method shall choose ^{131}I radioactivity isotope which features short half-life and low contamination, which can be used in a single borehole or multiple boreholes (in between) to test hydrogeological parameters.

3.8.3 The instruments shall conform to the following requirements:

1 The requirements for a gamma-ray tool are:

> 1) The spectral coverage of gamma-ray spectrograph used to measure low energy spectrum shall be adjustable with a spectrum stabilizer and its relative error shall be less than 1% for 8h of continuous reading.
> 2) When a radiometer is used for integral measurement, it shall be capable of measuring tens to hundreds of kev low-energy gamma-rays.
> 3) The sensitivity of a radiometer is high. When the

natural background is not greater than 0.72 PC/(kg·s) (10μR/h), the sensitivity threshold is not larger than 0.143 PC/(kg·s) (2μR/h).

4) The radiometer shows linearity within 0 PC to 3.6 PC/(kg·s) (0μR/h to 50μR/h).

2 The requirements for an environment gamma-ray tool are:

1) Measuring range.

The low measuring range is 1×10^{-8} Gy/h to 1×10^{-5} Gy/h;

The high measuring range is 1×10^{-5} Gy/h to 1×10^{-2} Gy/h.

2) Relative inherent error is less than 15%.

3) Energy response is 500kev to 3Mev and relative response difference is less than 30% (relative ^{137}Cs is referred to gamma-ray source).

4) Angular response is 0° to 180° and $\overline{R}/R$ is not less than 0.8 (For relative ^{137}Cs, refer to gamma-ray source. $\overline{R}$ is the mean value of angular response, and R the response value in the calibration direction).

3 The requirements for an alpha-ray tool are:

1) The large scintillator is used to make a scintillation detector whose efficiency A_m^{241} in detecting rays shall be greater than 60%.

2) The relative error between the readings under the maximum conditions and the normal readings shall be less than 15%.

3) Provided there is no adjustment in one week, the relative error of repeated readings shall be less than 15%.

4 The measuring apparatus for environment radon

concentration shall conform to the provisions of GB/T 14582.

  5 The requirements for an isotopic tracer instrument are:

    1) When V_v is greater than 0.1m/d, the relative error for testing vertical flow velocities shall be less than 2%.

    2) When V_f is greater than 0.05m/d, the relative error for testing horizontal flow velocities shall be less than 3%.

3.8.4 The layout of survey lines shall meet both the provisions of 4.1.3 in this code and the following requirements:

  1 The requirements for measurement of gamma-rays and alpha-rays are:

    1) The density of a line grid may be defined through the experiment. The measurements with a variety of accuracies are conducted on the known terrains, from which the data obtained are compared to the known geological ones to determine the optimum density of the line grid.

    2) The line spacing shall be 2cm to 4cm at the layout with no less than 3 survey lines passing through the major surveyed object. The station spacing shall be 0.5cm to 2cm at the layout, with the actual spacing of preferably 5m to 10m.

  2 In measuring the radon concentration in environment soil, the survey lines shall be laid out in terms of the geological structures and locations of faulted zones and they are preferably normal to the structural trend, their spacing being 10m to 20m and the station spacing 5m to 10m.

  3 In measuring the radon concentration in environmental air, the survey lines shall be laid out on the typical sites based on the building features, with the station spacing of 20m to 50m.

4 For the isotopic tracer method, one or multiple boreholes are deployed or chosen on a variety of hydrogeologic units, and the locations and depths of those boreholes shall meet the provisions of GB/T 50027.

3.8.5 The field measurement shall meet the following requirements:

    1 The requirements for measurement of gamma-rays are:

        1) Checking the grid density through knowing geological structures and springs inside or near the survey area, testing the measurement approach and optimal spectral coverage, finding out the magnitude of a normal field and the amplitude of an anomalous field.

        2) Before and after one day's work, the normal background value of an instrument shall be checked at a fixed site and the instrumental sensitivity inspected with the reference source.

        3) The same instrument shall be used by one person for the whole survey at the same survey line.

        4) The check shall focus on choosing of anomaly points and questionable stations, keeping an eye on various survey lines.

        5) Such artificial interferences as building and so on at a survey line shall be avoided to maintain the uniform measuring conditions at all stations.

        6) In conducting the environment gamma-ray measurement, the height of its probe above the ground surface is preferably 1m.

    2 The requirements for measuring alpha-rays are:

1) Alpha-ray has basically the same measuring requirements as gamma-ray. However the sampling is required at a station.
2) The choice includes taking soil samples or burying a static electric alpha-card.
3) When taking a static electric alpha-card sample, its station shall be chosen in the overburden in which the small pit for the buried card is flat at bottom without clay soil, broken stones, etc. The alpha-card burial duration may depend on the radiation field intensity and shall be greater than 4 hours. The time interval of taking the card for measurement shall be equal to that of the card burial.

3 While conducting the simultaneous measurement of gamma-rays and alpha-rays, the depth for measuring gamma-rays shall be kept in conformity with that of the pit from which a soil sample is taken for the alpha-ray testing.

4 The requirements for measuring environment radon concentration are:
1) The measurement of indoor radioactivity gas shall meet the provisions of GB/T 14582.
2) The probe shall be 1.5m high above the ground surface in collection of indoor radon and its daughter samples.
3) The regular and irregular test methods are preferred to measure the radon concentration in underground cavities and powerhouse air, and the monitoring frequencies and monitoring stations shall be identified as per the provisions of GB/T 16356.

5 The requirements for the isotopic tracer method are:
1) Before run in hole each time, the tool shall have its

surface background and its background value with a source loaded to be documented.

2) When testing hydrological parameters, appropriate radioactivity isotopes shall be chosen. ^{131}I is preferred to test groundwater flow velocity and direction, the dose for each time being less than 1×10^8 Bq.

3) The single-hole dilution method shall be used to test permeability velocity and flow direction, the single-hole tracer technique employed to test vertical flow velocity in the borehole and the multi-hole tracer technique chosen to test formation porosity.

4) The groundwater flow velocity and direction shall be measured in the hydrogeological observation hole, whose placement shall conform to the provisions of GB/T 50027, and the slotted casing shall be set in the Quaternary overlying strata, karst and the hole interval with developed structural fractures.

5) Such essential parameters as quantity of seepage, volume of saturated layers and porosity between the release position and station shall be estimated before conducting the experiment on interconnection of the seepage paths. For the multi-hole interconnection experiment, a large volume of radioactivity isotope needs to be released with ^{131}I used, completed at a time, and the maximum dose of release volume shall be less than 100×10^9 Bq each time.

6) After a release of isotope, one set of 5 to 10 readings shall be recorded every other 5min to 10min, and 5 to 10 sets of readings measured at each station.

7) The multiple-station isotope measurements shall be

made from deep to shallow. When there are more water-bearing layers, the packer in a borehole shall be used for a zonation test.

3.8.6 The check and assessment of data shall meet both the provisions of 3.1.7 in this code and the requirements for statistical fluctuation errors in radioactivity measurement:

   1 The statistical fluctuation errors of radioactivity shall be calculated according to the equation in Annex C.5 of this code.

   2 The standard fluctuation error σ shall be less than 30% and the relative standard error δ less than 10%, for the pulse counter.

   3 The standard fluctuation error σ shall be less than 20% and the relative standard error δ less than 10%, for the radiometer.

3.8.7 Data processing and interpretation shall meet the following requirements:

   1 The integrated analysis shall be made of the available geological data, geophysical exploration results, topography, terrains and climatic conditions to know the distribution and signatures of anomalies, and to discriminate their properties provided those spurious ones are eliminated.

   2 The anomalous point can be defined where the alpha-ray or gamma-ray intensity is over 3 times higher than the background value. If the overburden is thick, shielded by water body or affected by structures, and although their intensity is not up to over 3 times the background value, it is over 1.5 times that, it can still be identified as an anomaly.

   3 For the isotope tracer measurement in a borehole or in between holes, the permeability coefficient at each station shall be calculated according to equation in Annex C.5 of this code.

4 The effective dose equivalent is used as the evaluation criteria for the environment gamma-rays, the measured gamma-ray exposure rate shall be converted into the yearly effective dose equivalent and estimated according to the equations in GB/T 14583.

5 The equilibrium-equivalent (radon) concentration can be used as the evaluation criteria for the environment radon and its daughter concentration.

3.8.8 The result plots shall meet both the provisions of 3.1.9 in this code and the following requirements:

1 The plots for those measurements of gamma-ray, alpha-ray and environment radon concentration may include profiles and maps.

2 The plots associated with the isotope tracer method may include the location map of sampling points for water resource investigation, distribution map of seepage velocity with elevation, and the distribution map of groundwater flow velocity vector.

3.9 Comprehensive Logging

3.9.1 The comprehensive logging includes such methods as the electrical, acoustic, borehole seismic, radioactivity, temperature, electromagnetic wave or radar, fluid, magnetic susceptibility, sidewall ultrasonic imaging, borehole TV, caliper, inclination survey, etc.

3.9.2 The application conditions shall meet the following requirements:

1 The electrical logging shall be performed in the hole containing well fluid without metal casing. There shall be an electrical contrast between the target layer and adjacent layers, and also the target layer shall have given thickness.

2 The acoustic and seismic loggings shall be conducted in the hole with well fluid and without metal casing. There shall be an elastic wave velocity contrast between the target layer and adjacent layers, and also the target layer shall have given thickness.

3 The radioactivity logging can be used whether there is casing and well fluid or not. When the natural gamma logging is used for layering, there shall be natural radiation contrasts in any layers. When the gamma-gamma logging is used for layering, there shall be density variations in all layers.

4 The borehole TV shall be performed in the borehole which is a dry hole or a clean water hole, no casing being set.

5 The ultrasonic imaging logging shall be performed in the borehole containing well fluid without metal casing.

6 The fluid logging shall be performed in the borehole with well fluid or screen pipe and without metal casing.

7 The electromagnetic wave or radar and magnetic susceptibility logging shall be performed in the borehole without metal casing, and also the surveyed object shall be of a given dimension and present an electromagnetic contrast to its surrounding media.

3.9.3 The equipments shall conform to the following requirements:

1 The relative error for the depth counter shall be less than 2‰.

2 The requirements for tool insulation are:
    1) The insulation resistance shall be greater than $10M\Omega$ between the surface instrumental circuits, the instrument and ground, the draw works and ground,

the power supply source and ground.

    2) The insulation resistance shall be greater than 2MΩ between the cable core and ground, the electrodes, the downhole tool and its circuit housing.

3 The accuracy requirements for logging equipments are:

    1) The measuring error for potential difference is less than 3%.

    2) The measuring error for resistivity is less than 4%.

    3) The measuring error for caliper is less than 5mm.

    4) The measuring error for temperature is less than 0.5℃, and the thermal inertia is less than 3s.

    5) For the inclination survey, when the vertex angle of a borehole is greater than 5°, its measuring error shall be less than 0.5°, and the measuring error for azimuth less than 5°.

    6) Borehole seismic logging and acoustic logging equipments shall meet the provisions of 3.4.3 and 3.5.3 in this code, respectively.

4 The recording accuracy requirements for logging tools are:

    1) The line width of a curve recorded by analog logging instrument shall be less than 0.5mm and the wobble width of a record curve caused by the instrument's inherent noises shall be less than 1mm.

    2) The sampling interval for records of a digital logging tool is not less than 1 sampling point every 0.05m.

3.9.4 The measurement shall meet the following requirements:

    1 The requirements for marks of logging cable length are:

        1) Before used, the new cable shall hang a heavy object that is equal to the tool's weight and be run up and

down 5 times in a borehole, and then the fixed depth marks would not be done until its extension is stabilized.

2) The interval for a depth mark is 10m, and when 1 : 50 depth scales is used to make detailed survey, that is preferably 5m and the error in length shall be less than 2‰.

3) It shall be checked per year or per 10 wells.

2 Before logging in a borehole, a heavy hammer, which is equivalent to the diameter and length of a downhole tool, may be used for exploring the hole.

3 The depth scale of logging data or curves shall be in conformity with that of a borehole columnar section and the same depth scale be used in the same survey area. For the hole interval needed for detailed survey, additional large-scale auxiliary records are required.

4 The lateral scale shall be determined based on geological data and trial logging data or curves and the large scale is chosen depending on the non-exceeding values of most of curve records.

5 The depths for raw logging data or curves shall be accurately marked and recorded:

1) When logging for a separate trip or interval, the repeated observation of at least one depth mark shall be done at the joint of key data or curves.

2) The position of the zero line in a log, if any, shall be recorded at the head and end of data or a curve. If there is no zero line in the log, the reference baseline for lateral coordinate shall be marked at the head and end of the curve.

6 The probe moving speed of all loggings, whether it is up or down, shall remain constant and not exceed the restricted speed requirement in Table 3.9.4.

7 the logging of temperature, fluid resistivity and borehole TV shall be run with the probe moving down the borehole, other loggings with the probe moving up the borehole to reduce the depth errors.

Table 3.9.4 Speed limits for lowering and
lifting a logging cable   Unit: m/min

Logging methods	Depth scale		
	1:200	1:100	1:50
Electrical logging (exclusive of micro-electrode system)	20	10	5
Micro-electrode, caliper	10	6	3
Acoustic, radioactivity, temperature, electromagnetic wave logs	5	3	2
Borehole TV observation, ultrasonic imaging	The image shall be clear		

3.9.5 The onsite logging jobs shall conform to the following requirements:

1 The electrode spacing for electrical logging shall be identified on experiment in terms of the logging tasks and conditions in different survey areas. The current logging is required to reduce circuit resistance and grounding resistance of surface electrodes and to ensure power supply at constant voltage, and while recording a current curve, it shall be checked and the increment direction shall be determined. The spontaneous potential logging is required to employ unpolarizable electrodes. When using an attached metal weight for easy access to the borehole, the measuring electrode shall be 2m away from the metal weight.

2 Before natural gamma ray logging, a reference source shall be employed to check the tool. The statistical fluctuation shall be recorded in the hole sections of argillaceous rock, its recording duration being 10 times the time constant chosen from the curves. For gamma-gamma logging, a density calibrator, if any, shall be used for wellsite calibration. If there is no density calibrator, an apparent density measurement should be made, in which the chosen source intensity should be high enough to suppress the interferences from natural gamma rays and it be 20 times larger than the average amplitude value of natural gamma rays in the primary layer of interest.

3 The depth scale for ultrasonic imaging shall be defined on the basis of dip of layers, cavities, fractures, size of faults, thickness of weak interlayer as well as accuracy of observation.

4 The borehole TV shall be employed to conduct tracing observation of the major geological anomalies and their images shall be largely distinct.

5 For the fluid log, the hole wall shall be cleaned out.

6 When using multiple logging methods, the temperature logging shall be run first.

7 Before and after a caliper log, the tool shall be calibrated on site, and there shall be over 3 calibration records for different diameters with their error less than 5mm.

8 Before a borehole inclination tool is run in hole, the magnetic compass or clinometer shall be used on site to make a simple measuring inspection and zeroing (the vertex angle is 0°) of vertex angles. The station spacing of an inclination survey shall be less than 5m, and when hole deviations change significantly (the vertex angle difference is greater than 2° and the azimuth difference is greater than 20°), the infilled stations shall be added.

9 For the electromagnetic wave or radar logging, a frequency or multi-frequency can be chosen for measuring in terms of the geophysical conditions and dimension of a surveyed object.

3.9.6 The check observation shall conform to the following requirements:

1 The check quantity for a inclination survey shall be greater than 20% and that of other logging methods shall be more than 10%. When a special anomaly is found from some of logging data or curves, the corresponding hole segment shall be repeatedly run.

2 The repeatedly run curves shall be in substantial agreement with the raw logging curves.

3.9.7 The check and assessment of data shall meet the following requirements:

1 The assessment of data is divided into two categories: qualified and unqualified.

2 The unqualified exists in one of the following situations:
1) More than 2 depth marks are continuously missed in the target segment.
2) There are more than 5% discontinuities, missing or distortions in a curve.
3) The records at the head and end of a curve, up and down speeds, depth scales, lateral scale, and check quantity fail to meet the requirements in this code.

3.9.8 Data processing and interpretation shall meet the following requirements:

1 The logging data need to be edited, processed and interpreted, and digitized.

2 Compared analysis of different parameter curves shall be made according to the layering characteristics in various logging curves and the layering can be conducted combining the geological and drilling information and in terms of the physical properties and geological names to determine the depth and thickness of layers or geological mass.

3 When there are the uniform geological conditions in the same survey area, the unified interpretation shall be done especially for the anomalies in logging curves.

4 The true formation resistivity shall be measured with lateral logging (transverse electrical sounding), and when the lateral logging conditions are not available, the true formation resistivity in relatively thick layers can be determined according to resistivity curves and by taking into account such parameters as caliper, fluid resistivity, host rock resistivity, etc. For the lateral logging that is available, it can be obtained from lateral logging curves after being corrected.

5 When conducting the interpretation and inference of data from acoustic logging and gamma-gamma logging, the whole borehole shall be divided into several thick layers with different velocities and densities at a macro-scale to acquire their mean wave velocity and mean density values. Then, the geological inference and demarcation of thin layers can be done through analysis of various anomalies. In the layers where their thickness is less than 30cm, the approximate range of their wave velocities or densities shall be calculated based on logging curves.

6 Characterization of geological signatures in a borehole shall be given with the borehole televiewer or ultrasonic imaging, calculating the dip, trend and thickness of fractures, faults and soft interlayers, etc. Provided the vertex angle in an inclined hole is greater than 5°, the inclination angle must be

corrected utilizing the data of caliper and inclination survey in calculating occurrence.

7 When measuring the inclination in an inclined hole, the projection shall be mapped in the horizontal plane and vertical plane of the borehole.

3.9.9 In addition to the provisions of 3.1.9 in this code, the result plots shall also meet the following requirements:

1 In the synergistic interpretation chart, the allowable depth errors in this code shall be adjusted within the adjacent depth marks. Each adjustment at one point shall be less than 1mm.

2 The depth coordinates for all curves shall be uniform in the same plot.

3 The curves, as those from electrical, acoustic, radioactivity, temperature and caliper logging, shall be mapped in a result plot for comprehensive logging. The ultrasonic imaging log shall be affixed to the borehole logging columnar section of the plot. The data from other logging methods can be individually plotted or listed. Their results, however, shall all be reflected in the interpretation plot for comprehensive logging curves in text format.

4 If there are many boreholes in a geologic cross-section or dense distribution of the holes in a survey area, the geological interpretation profile across the entire cross-section shall be mapped on the basis of the single-hole logging interpretation, in which a few of logging curves can be chosen, which are representative and clearly reflected to the layer of interest and for ease of comparison.

5 For observation of a borehole televiewer, its plots that shall be submitted include the edited images and the images with typical geological signatures.

4 Comprehensive Applications of Geophysical Methods

4.1 Investigation of Overburden

4.1.1 A variety of techniques, including electrical survey, seismic exploration, hydro-acoustic survey, ground penetrating radar, elastic wave testing and comprehensive logging, are available for use in exploration operations such as determining the thickness of each overlying layer and the topographic relief of bedrock top plates, measuring the physical parameters of each overlying layer, identifying the topography of paleo-channels or bedrock valleys, and performing the layering and assessment of natural building materials.

4.1.2 Investigation methods and techniques shall meet the following requirements:

1 Survey grids and lines shall be laid out as required by the provisions of 3.1.3 in this code.

2 When the conditions specified in the regulations of 1, 2 or 7 in the provisions of 3.1.3 in this code are met between overburden and bedrock, the electric sounding or CSAMT can be used.

3 When the conditions specified in the provisions of 3.3.2 in this code are met between overburden and underlying bedrock, the ground penetrating radar can be used for investigation.

4 When the conditions specified in the regulations of 1 or 2 in the provisions of 3.4.2 in this code are met between overburden and underlying bedrock, the shallow seismic refraction method can be used.

5 When the conditions specified in the regulations of 1 or 3 in the provision of 3.4.2 in this code are met between overburden and underlying bedrock, the shallow seismic reflection method can be used, in combination with other geophysical exploration techniques or drilling which will allow contrast analysis to be done.

6 When relatively thin layering is required for the overburden and the investigation depth is small, and the conditions specified in the regulations of 4 in the provision of 3.4.2 in this code are met, the Rayleigh wave method can be used, preferably in combination with other geophysical exploration techniques or drilling which will allow contrast analysis to be done.

7 In the water areas that have a wide river surface, a large water depth, a slow water flow and small sedimentary particles, such as those in reservoirs, lakes, shallow sea areas, ports and wharfs, sonic echo exploration or seismic imaging can be used.

8 For the main survey lines and geologically complex terrains, comprehensive geophysical exploration techniques should be used.

9 Comprehensive logging or elastic wave testing can be used for measuring physical parameters of all overburden and layering them.

4.1.3 Data processing and interpretation shall meet the following requirements:

1 Analysis and interpretation shall be done based on differences in velocity, resistivity and density between the overburden and underlying bedrock.

2 The relationship between physical property and geology in a layer shall be analyzed using data of physical properties in boreholes and outcrops, and near-hole electrical sounding and seismic profile data. In case of inconsistencies occurring between the property and geology in the layers, the physical thickness shall be used for determining physical parameters, and this shall be indicated in the result report.

3 Physical parameters shall be analyzed for their horizontal variations, and shall be used correctly to make a quantitative interpretation.

4 It shall be pointed out whether the interpreted thickness of the overburden includes that of the strongly weathered zone belt in the bedrock.

5 In addition to the requirements for the provisions in 3.1.9 in this code, the plots of results shall also contain overburden isopach maps and bedrock contour maps. When there exist conditons that permit layering of the overburdens, the isopach maps and contour maps of a layer can be plotted.

4.1.4 Survey accuracy shall meet the following requirements:

1 In case of favorable geophysical conditions and available horeholes information, relative errors in depth shall be less than 15% for overburden with a thickness of over 10m, and shall be less than 20% for poor geologically survey areas.

2 When sonic echo exploration is used for a water depth of less than 50m, relative errors in bedrock buried depth shall be less than 15%.

4.2 Investigation of Buried Geological Structures and Fractured Zones

4.2.1 Techniques that are available for investigating the position, dimension and extension of buried geological structures and fractured zone, and measuring physical parameters of structural fractured zones, include electrical survey, GPR, seismic survey, CT, radioactivity survey and comprehensive logging.

4.2.2 Investigation methods and techniques shall meet the following requirements:

    1 In addition to the requirements for the provisions in 3.1.3 in this code, survey grids should have a station spacing less than half the width of a structural fractured zone.

    2 For the buried geological structures and fractured zones in a concealed structure that displays a low-resistivity anomaly, the transient electromagnetic method, CSAMT, electrical profiling, electrical sounding and resistivity imaging can be used.

    3 For the tectonic fractured zones enriched with ground water, a combination of induced polarization and electrical sounding techniques is preferred. When groundwater activity exists, and an electrofiltration field is generated which leads to a significant self-potential anomaly, the self-potential method can be employed.

    4 When the structural fractured zones are exposed in borehole and show good electrical conductance, the mise-a-la-masse method can be used to determine their strikes.

    5 For the overburden with a small thickness (buried depth less than 10m) and a high resistivity, GPR can be used to explore structural fractured zones.

    6 For the moderate width of the geological structures and fractured zones, the shallow seismic refraction method can be used, where amplitude comparison should be made based on the same gain level and scanning size, preferably with a station spacing of 5m. For determining the position of the fault fractured zones with a smaller width mainly based on the dynamic properties of waves, the station spacing should be 2m.

    7 For faults with a vertical fault throw and a moderately large elevation difference between their hanging walls and footwalls, the shallow seismic reflection method can be used in combination with an observation system that depends on horizontal multiple stacking or common offset profile, preferably with a station spacing of 2m to 4m.

8 For fault fractured zones that show good gas and water permeability, with radioactive gas going up the faulted zones to the surface, the radioactivity survey can be used.

9 When boreholes are available, comprehensive logging can be used to measure the physical parameters of the structural fractured zones and their distribution range along the axis of a drill borehole, whereas cross-hole CT can be employed to investigate the position, dimension and extension of structural fractured zones.

4.2.3 Data processing and interpretation shall meet the following requirements:

1 Comprehensive analysis and interpretation shall be performed based on various anomalies in the physical properties of the structural fractured zones.

2 The spatial distribution of fault fractured zones should be determined by data collected from comprehensive geophysical exploration and in combination with the geological characteristics of the survey area.

3 For combined profiling parallel array method, low resistivity orthogonal-point anomaly may be used to infer the position and strike of fault and fractured zones.

4 For the TEM, CSAMT and resistivity imaging, ρ_s cross section map can be used to infer the position, dimension, extension, trend and approximate dip angle and dip direction of fault fractured zones.

5 For induced polarization, any interference effects caused by carbonaceous strata and metal ore mass shall be ruled out.

6 For GPR data, anomalies that show up in reflection series can be used to infer the position of fault and fractured zones.

7 For the shallow seismic refraction method, sections where an sharp increase or a break in the slope of $\theta(x)$ curve, or a more than 30% decrease in the interface velocity is detected may be interpreted as low velocity zones. When some patterns, such as a decreasing amplitude, increasing apparent cycles and waveform anomalies, are found in gross seismic records on the low velocity zones, it may be inferred that the zones are structural fractured ones. The width of low velocity zones may be used to infer the approximate width of fractured zones. For small fault fractured zones, the features of low velocity zones may often be vague, and thus the position and width of fractured zones shall be determined by analyzing waveform changes, amplitude damping and wave displacement in data obtained from short-geophone-interval investigation.

8 For the shallow seismic reflection method, the comparing and analyzing work shall be done in common depth point stacking-time profile or common offset profile. The position of faults may be deducted based on such features as breaking the event of the marker layer in the bedrock, decreasing amplitude of reflected wave and wave displacement. The approximate throw of faults may be determined using event break time difference.

9 For a smaller overburden thickness (with a buried depth of less than 10m), the position of fault fractured zones may be inferred from the fact that physical anomalies at the ground surface caused by natural electrical fields, induced polarization parameters and radioactive intensity are higher than 1.5 times that of normal fields, and are distributed in stripe patterns on a plane.

10 In addition to the provisions of 3.1.9 in this code, the plots shall also contain the distribution plan of an underlying structural fractured zone.

4.2.4 Survey accuracy shall meet the following requirements:

1 If favorable topography and geophysical exploration conditions exist in survey areas, any fault fractured zones with a width of more than 2m shall be able to be detected for an overburden thickness of less than 10m, whereas for a thichness of greater than 10m, fault fractured zones at a width which is one quarter of the thickness of overburden shall be able to be identified. For fault fractured zones with a dip angle of less than 45°, the dip direction and approximate dip angle shall be determined.

2 CT-based investigations shall be able to identify any fault fractured zone at a width which is greater than one tenth of the hole-to-hole spacing.

4.3 Investigation of Karst

4.3.1 Such techniques as electrical survey, GPR, seismic survey, CT and comprehensive logging may be chosen for investigation of karsts in terms of their distribution, buried depth, dimension, extension direction, infilling and groundwater.

4.3.2 Investigation methods and techniques should meet the following requirements:

1 When bedrock is exposed, GPR, transient electromagnetic method and CSAMT can be used to find the distribution, dimension, extension direction and infilling of karst.

2 For thin overburden, electrical profiling, resistivity imaging and shallow seismic refraction methods can be used to investigate karst channeling furrow, fluid bowls and soil caves, while resistivity imaging, transient electromagnetic method, seismic reflection method and Rayleight wave method can be used to identify low-buried karst cavities and structures.

    3 For thick overburden, CSAMT can be used to reveal the distribution, dimension and trend of karst.

    4 GPR can be used to determine the distribution of karst around tunnels.

    5 CT scanning can be used to explore karst cavities deeply buried and their infillings between holes.

    6 Comprehensive logging can be used to investigate the corrosion of karst strata at hole walls, the position of drain openings of karst underground streams or springs and karst water table, and reveal small karst cavities through observation drill boreholes.

    7 To investigate building foundations where space is limited, techniques such as GPR, resistivity imaging and seismic imaging are preferred.

4.3.3 Data processing and interpretation shall meet the following requirements:

    1 Data obtained using electrical profiling, transient electromagnetic method, electromagnetic wave CT and CSAMT should be interpreted based on such anomaly characteristics as the infilled karst cavities display low resistivity and high absorption, while unfilled karst cavities often display high resistivity and low absorption.

    2 The interpretation of the shallow seismic refraction method and elastic wave CT should be based on such properties as low velocity and low penetrating energy of karst cavities.

    3 The distribution, dimension and top buried depth of

karsts should be determined for the shallow seismic reflection method by the features of strong reflection generated from the top of karst cavities while for GPR by the hyperbolic reflections from karst cavities.

4 Differences in anomalies between electromagnetic and elastic wave data should be analyzed and compared to offer an understanding of how karst caves are infilled. Induced polarization parameter anomalies can also be used to find the properties of infilling materials.

5 Logging data should be interpreted based on karst characteristics including temperature, electrical properties, wave velocity, density and television image features.

6 The connectivity, extension directions, and range of karst cavities should be inferred and interpreted based on anomalous positions of various profiles in the survey area, taking into account relevant hydrogeological records.

7 In addition to meeting the requirement of provisions in 3.1.9 in this code, the plots should also contain a karst distribution map.

4.3.4 Survey accuracy should meet the following requirements:

1 For cavities that are larger than one tenth of the hole spacing and the effective rate of cross-hole CT investigations shall be more than 80% effective.

2 For cavities that are larger than one tenth of the buried depth, the effective rate of surface geophysical exploration of karst cavities shall be more than 70% effective under thinner overburdens.

4.4 Investigation of Thickness of Weathered Rock Mass and Depth of Unloading Zones

4.4.1 Techniques that may be used to measure rock mass

weathered zone thickness, make weathering zoning and assess weathering degree, and determine side slope unloading zone depth and the scope of its effects include electrical survey, GPR, seismic survey, CT, elastic wave testing and comprehensive logging.

4.4.2 Investigation methods and techniques should meet the following requirements:

1 Survey grids shall be laid out in the way that the requirements specified in the provisions of 3.1.3 in this code can be met. In addition, survey lines for the investigation of rock mass weathered zone thickness shall be arranged in the same way as those for the investigation of buried structural fractured zones. Survey lines for measuring unloading zone depth shall be arranged mostly perpendicularly to side slopes, and a few auxiliary survey lines which are parallel to side slopes should also be laid out. For areas that show a significant width variability, more survey lines and points shall be added as appropriately, and survey lines shall be long enough to reach sections of the intact rock mass where no unloading is detected.

2 When bedrock surfaces are not covered by any overburdens or they are covered just by thin overburdens, GPR can be used.

3 If bedrock surfaces are covered by moderately thick overburden, weathering boundary surfaces have up-and-down (uneven) topography, and weathered zones vary widely in physical properties, the shallow seismic refraction method, electrical sounding and resistivity imaging may be used.

4 When overburdens and weathered zones are both thick, the shallow seismic reflection or refraction method may be used.

5 If a detailed investigation is needed for rock masses with irregular local weathering features, CT may be used.

6 When boreholes or shafts are available, the comprehensive logging or elastic wave testing should be used.

7 When rock mass weathered zone thickness and unloading zone depth investigations are performed in adits or shaft, the single-hole acoustic wave, penetrating acoustic wave or continuous seismic wave testing may be used. Testing holes shall be placed at the same height on adit walls and at the same plane, and a down declivity should be 5° to 10°.

4.4.3 Data processing and interpretation shall meet the following requirements:

1 Analysis and interpretations shall be based on the differences in wave velocity, resistivity and density between rock mass weathered zones and unloading zones.

2 The relationships between the significantly sharp changes at the abnormal points in physical property curves with lithology, fault fractured zones, the position of the rock mass weathered and unloading interface shall be determined. The characteristics of changes in weathered and unloading zones shall be known.

3 When investigating the rock mass weathered zones thickness and unloading zones depth, the distributional relationship among the formation lithological interfaces, fault fractured zones and unloading zones shall be found out.

4 For weathered zones arranged in a layered pattern, the thickness of each layer in the layered weathered zones may be determined. If a suitably large number of boreholes and test pits exist in the survey area, contrast analysis shall be performed for test results from these pits and boreholes to determine the interfaces of each weathered zone.

5 In case of weathered zones showing a gradual gentle continuous velocity change in the vertical direction, the relationship between the velocity and depth shall be identified.

6 The classification of rock mass weathered zones based on P-wave velocity ratio between weathered and unweathered rock masses shall conform to GB/T 50287 requirements.

7 In addition to meeting the requirements of 3.1.9 in this code, the plots of results shall also contain rock mass weathered zones contour maps or isopach maps, and maps of range of unloading effect. Velocity in unloading rock masses as well as in non-unloading ones shall also be indicated.

4.4.4 Depth accuracy shall meet the following requirements:

1 If just a small number of elastic wave testing data from boreholes, adits, or test pits are available, the relative error in depth and thickness shall be less than 20% for making layering of subsurface rock mass weathered zones.

2 When no borehole data is available, or in the case of unfavorable topographic, bad geological or physical conditions, the relative error in depth shall be less than 30%.

4.5 Investigation of Weak Interlayers

4.5.1 To determine the position, thickness and physical parameters of weak interlayer, the comprehensive logging and elastic wave testing may be used.

4.5.2 Investigation methods and techniques shall meet the following requirements:

1 If weak interlayer in borehole is less than 20cm in thickness, large depth scale shall be used.

2 For the investigation of weak interlayer in borehole within sandy gravel overburden, the following requirements shall be met:

1) For mud-protected borehole, the natural gamma ray

logging, lateral or apparent resistivity logging or self-potential logging should be used. For sandy gravel beds where ground water has a high seepage velocity, the diffusion process in the well fluid resistivity logging should be used. When significant density and acoustic velocity differences exist between interlayer and sandy gravel beds, the gamma-gamma logging, acoustic logging and lateral ultrasonic imaging may be used.

2) Electrical logging should not be selected when SM vegetable gum is used in drilling operations.

3) For cased boreholes, the natural gamma logging and gamma-gamma logging shall be used.

3 Requirements for the investigation of weak interlayer in bedrock borehole are:

1) Ultrasonic imaging, natural gamma logging and self-potential logging shall be selected by testing based on geophysical conditions in the survey areas and the differences in physical properties between weak interlayer and host rock.

2) If the investigated sections have no casing but contain well fluid, lateral or apparent resistivity logging, caliper logging, acoustic logging or gamma-gamma logging may be selected.

3) For clear well fluid, borehole TV may be used for observation.

4) When the investigated sections are in dry boreholes without any casing, the gamma-gamma logging, caliper logging and borehole TV may be used, which is preferably complemented by natural gamma log-

ging.

    5) When the investigated sections are fitted with casing, the natural gamma logging or gamma-gamma logging should be selected.

4 when physical parameters of weak interlayer are measured in adits, the acoustic method may be used.

4.5.3 Data processing and interpretation shall meet the following requirements:

  1 Analysis and interpretation shall be based on the unique physical parameters of weak interlayer including low wave velocity, low resistivity, low density, high yield value and high natural gamma radiation intensity.

  2 Contrast analysis shall be performed for a variety of logging data. When a large number of boreholes need to be investigated in the survey area, contrast analysis shall be done for the changes of depth and thickness in each borehole.

  3 The dynamic parameters of weak interlayer may be calculated from the velocity and density of compression and transverse acoustic waves.

  4 Borehole TV may be used to examine weak interlayer for their depth, thickness and occurrence state.

  5 The result plots shall contain comprehensive borehole logging result plots and tables that list weak interlayer statistics including depth, thickness and physical parameters.

4.5.4 Survey accuracy shall meet the following requirements:

  1 For boreholes whose diameters are less than 100mm, walls are relatively intact, and the dipsof weak interlayer are less than 30°, survey accuracy shall meet the following requirements:

    1) For borehole TV-based investigation of weak interlayer,

any interlayer thicker than 1mm shall not be missed. The absolute error in determining the thickness of interlayer shall be less than 3mm for the interlayer with a thickness of below 20mm.

2) Micro-electrode logging and ultrasonic imaging logging, when used for the investigation of soft interlayer, shall not ignore interlayer more than 5cm in thickness.

3) For the investigation of soft interlayer using natural gamma logging, self-potential logging, apparent resistivity logging, acoustic logging and gamma-gamma logging, interlayer thicker than 10cm shall not be missed.

2 The relative error of depth in the soft interlayer investigation results shall be less than 5‰.

3 The absolute error of angle of dip of soft interlayer obtained from borehole TV observations or ultrasonic imaging investigation shall be less than 5°, and the absolute error for direction of dip less than 10°.

4.6 Investigation of Landslide

4.6.1 Techniques such as seismic survey, electrical survey and comprehensive logging may be used for investigating the dimension and thickness of landslide and the topographic relief of sliding beds, landslide stratification, determining the distribution and buried depth of saturated zones and aquifers inside the landslide, identifying the properties of sliding zones and measuring the physical parameters of landslide.

4.6.2 Investigation methods and techniques shall meet the following requirements:

1 Grid or fan-shaped survey grid should be used, and survey lines should be aligned in the direction of principal sliding zones. In case of deploying along slopes or parallel contour lines, the survey lines should be extended beyond the boundary of landslide.

2 For landslides which are composed mainly of clay silt, sand and rubble soil, the shallow seismic refraction method, the shallow seismic reflection method and the Rayleigh wave method may be used.

3 When landslides mainly are large and loose collapse rock, techniques such as CSAMT, electrical sounding, resistivity imaging and shallow seismic refraction method may be used.

4 For the investigation of ground water and aquifers in landslide, the induced polarization, electrical sounding, resistivity imaging and CSAMT may be used.

5 When boreholes are available, the comprehensive logging may be used to determine the properties of slip surface and the physical parameters of landslide.

4.6.3 Data processing and interpretation shall meet the following requirements:

1 Analysis and interpretation shall be based on variations of landslide in wave velocity, density, resistivity and natural gamma intensity.

2 In case of electrical sounding curves distortions occurring due to changes in horizontal resistivity in landslide, data obtained from electrical survey should be analyzed in combination with data obtained from other investigation techniques.

3 The shallow seismic refraction method may be used to calculate the thickness of landslide. When velocity interfaces exist in landslide, multiple forward and reverse shooting time-distance curves shall be used for stratifying.

4 The shallow seismic reflection method shall be combined with the shallow seismic refraction method and well logging method to determine the events of the sliding surfaces and interface within the landslide.

5 The boundary of landslide may be determined based on overburden thickness, wave velocity and changes in apparent resistivity, in conjunction with topographic features.

6 In addition to meeting the requirements of 3.1.9 in this code, the plots of results shall also contain landslide isopach map and slip surface contour map.

4.6.4 The relative error in the determination of landslide thickness and depth shall be less than 15%, or 20% under complex conditions.

4.7 Detection of Hidden Defects in Dams and Embankments

4.7.1 Techniques such as electrical survey, GPR, elastic wave testing and isotope tracer technique may be used to detect various hidden defects in dams and embankments and earth rock dams, including cavity voids, cracks, soft zones (including soft embankment sections), sand beds (including sandy embankment sections) and leakage paths, and determine their dimension, position and buried depth.

4.7.2 Investigation methods and techniques shall meet the following requirements:

1 Before starting investigation, it is necessary to examine and record external appearance of embankments and dams and collect records on emergency events during each of the previous flood seasons.

2 At first, reconnaissance survey should be carried out. Then, based on the results, some representative sections should be selected for a detailed investigation, and these sections should have a total length larger than 20% of that of sections to undergo the reconnaissance survey, and shall cover all hidden defect types and include those anomalies whose properties are difficult to determine as well as a fraction of sections without anomaly.

3 When the width of the top of embankment is less than 4m, only one survey line needs to be placed along the abutment on the upstream side. When the width is greater than 4m, each side of the embankment should have a survey line. If further investigation is needed to reveal the exact conditions of hidden defects, more survey lines may be added on the slope or at the toes of embankments, or perpendicularly to the embankment axis.

4 For the investigation of cavity voids, cracks and weak zones in dry embankments, techniques such as resistivity imaging, electrical profiling, transient electromagnetic method and GPR should be used.

5 GPR and elastic wave testing may be used to investigate void areas in protection revetments.

6 To find hidden defects in embankment or dam foundations, CSAMT may be used.

7 Isotope tracer technique should be used to measure the seepage parameters of embankments or dams.

8 To investigate leakage paths lying under the lines of seepage water saturation line, self-potential method and isotope tracer technique should be used.

4.7.3 Data processing and interpretation shall meet the

following requirements:

1 Investigation data shall be analyzed and principles of interpretation shall be established in the light of the history of embankments or dams, the reinforcement measures that have been taken, and the flood-season emergency records.

2 Embankment sections with poor soil conditions may be defined according to the background values of resistivity.

3 The nature of various hidden defects, including voids, cracks, soft zones, high sand content zones and leakage, should be clearly identified.

4 Parameters such as the number of milestone and the buried depth of each hidden defect shall be provided to indicate its spatial position.

5 For a given profile or investigation technique, contrast analysis may be made for the differences in the position and amplitude of anomalies between dry seasons and rainy seasons, thereby determining the position and nature of leakage paths.

6 The position and flowing direction of seepage paths may be analyzed and determined based on the relationship between the natural electrical potential and the water flow.

7 Embankments or dams under investigation may be classified into three types of sections, namely good quality sections, relatively good ones and hidden defect growing ones, based on the distribution of hidden defects, the background values of resistivity and radar image features, taking into account the characteristics of hidden defects in embankments and dams.

8 The characteristics and spatial position of each type of hidden defect shall be clearly labeled in the main result plots submitted.

4.7.4 Survey accuracy shall meet the following requirements:

1 The type of hidden defects shall be more than 70% correct.

2 The absolute error of plan location of hidden defects shall be less than 2m.

3 The relative error in measuring the buried depth of hidden defects shall be less than 20%, and the absolute error less than 2m.

4.8 Advanced Prediction in Tunnel Excavation

4.8.1 For advanced prediction in tunnel excavation, in order to determine existence of any unfavorable geological bodies ahead of the working face, including fault fractured zones, karst cavities, ground water enriched areas, etc. such techniques as seismic negative apparent velocity method, elastic wave vertical reflection method and GPR may be chosen.

4.8.2 Investigation methods and techniques shall meet the following requirements:

1 If the prediction area is large, the working face can not be used for investigation, and target bodies have a moderately large reflection surface, seismic negative apparent velocity method may be used.

    1) Either extended spread type observation system or cross spread type observation system may be selected for investigation, based on local geophysical conditions. For the former system, the mode of one transmitter and multiple receivers or that of one receiver and multiple transmitters" may be used, with receiving points or explosive source points arranged on side walls or bottom surface behind the working face, aligned along the tunnel axis and

spaced at 2m to 5m intervals. Explosive sources in the one-transmitter-and-multiple-receiver mode or receiving points in the mode of "one receiver and multiple transmitters" shall be placed at the furthest position away from the working face. For the latter, a large number of detection points and source points shall be arranged at equal intervals (preferably at 2m to 5m) and symmetrically on both side walls in the tunnel section behind the working face, with one side used for shooting and the other for receiving in the reverse order.

  2) Geophones shall conform to technical requirements specified for the devices used, and shall be placed in holes created to accommodate them. For sources, in-hole explosion shall be used.

2 When the prediction area is small and the working face is available for investigation, GPR may be used.

  1) A given number of profiles shall be arranged on the working face which serves as a centre. Station measurements, if used, shall be performed at a station spacing of less than 0.5m.

  2) GPR antennae shall be selected based on specific conditions and investigation ranges.

3 When the prediction area is large and the working face is available, the elastic wave vertical reflection method may be used.

4 For complex tunnel section, a comprehensive prediction approaches should be used.

4.8.3 Data processing and interpretation shall meet the following requirements:

1 Analysis and interpretation shall be based on the fact that unfavorable geological bodies such as fault fractured zones, water or mud-filled karst cavities and ground water enriched zones tend to exhibit unusual physical properties such as low wave velocity and low resistivity, and that there are significant differences in physical properties between unfavorable geological bodies and surrounding rock.

2 For the seismic negative apparent velocity method, the T—D curves respectively for positive and negative apparent velocity shall be extended by extrapolation until they intersect at a point, and based on the observation that the T—D curve for incident wave displays positive velocity while the T—D curve for reflected wave shows a negative one, this intersection point can be interpreted as the prediction interface position of the unfavorable bodies.

3 For the elastic wave vertical reflection method, the features of reflected waveform on time profiles shall be relied on to infer and predict whether or not unfavorable geological bodies exist, whereas reflection times and the velocity parameters of surrounding rock at the working face shall be used to determine the location of the unfavorable bodies.

4 When GPR is used, the position and nature of unfavorable geological bodies shall be determined and predicted based on observed radar image features, such as abnormal shapes, waveform patterns, and electromagnetic wave attenuation characteristics.

5 For geologically complex tunnels, in addition to carefully analyzing geophysical exploration results, it is also necessary to get a detailed understanding of the geology at the surface, taking account of the geological records and physical parameters

obtained from current and previous excavation operations.

4.8.4 When geophysical conditions are favorable and the construction work is performed without any serious disruption, the comprehensive geophysical exploration method, when applied to advanced prediction in tunnel excavation, shall be more than 80% correct.

4.9　Investigation of Groundwater

4.9.1 Techniques such as electrical survey, radioactivity survey, comprehensive logging and seismic investigation, may be used in identifying aquifers and impermeable layers in the Quaternary formation and measuring their depth and thickness, determining the water abundance of bedrock fractured zones, karst development zones and fault fractured zones, performing geothermal water investigation, measuring ground water tables, establishing fresh-to-salt water interface, and delineating the polluted ground water areas and monitoring the pollution of ground water.

4.9.2 Investigation methods and techniques shall meet the following requirements:

1 Ground water investigation shall be performed in close coordination with hydrogeological survey, drilling and testing operations.

2 For investigation of the area of ground water, the survey grid should be laid out in a net-shaped configuration. For areas where a good understanding has been obtained regarding their hydrogeological conditions, the arrangement of survey lines may be limited to important sections and questionable sections.

3 When boreholes are available, the comprehensive logging shall be used.

4 For the investigation of ground water in the overburden and underlying bedrock, the first step is to make a detailed survey targeted at aquifer (zones), in order to make clear how water-abundant these layers (zones) are:

    1) The investigation of aquifers in the Quaternary formation shall be targeted at water-bearing sandy gravel layers, and seismic survey and electrical survey may be applied to the investigation of the layers of interest. To measure the water abundance in the layers of interest, techniques including electrical sounding, induced polarization, transient electromagnetic method and CSAMT may be used.

    2) The investigation of aquifers (zones) shall be targeted at fractured zones, karst development zones and fault fractured zones in the formation, and seismic survey and electrical survey may be used to investigate the layers of interest. To measure water content in the layers of interest, techniques such as electrical sounding, induced polarization, self-potential method, transient electromagnetic method, CSAMT and radioactivity survey may be used.

5 Induced polarization, electrical sounding, transient electromagnetic method and CSAMT may be used to determine ground water tables.

6 To determine the flow velocity, flow direction and seepage velocity, self-potential method and mise-a-la-masse method may be used. Where boreholes are available, the isotope tracer technique is preferred.

7 For the investigation of ground water and geothermal

sources that are buried at a large depth, transient electromagnetic method or CSAMT may be used.

8 To determine the distribution of the Quaternary underground salt water and fresh water on the horizontal plane, and to delineate the polluted ground water areas and monitor the pollution of ground water, electrical sounding, transient electromagnetic method and CSAMT may be used, whereas comprehensive logging shall be used for determining salt-to-fresh water interfaces in multi-layer underground aquifers.

4.9.3 Data processing and interpretation shall meet the following requirements:

1 It is necessary to establish a number of typical abnormal geophysical curves under typical hydrogeological conditions based on hydrogeological survey and drilling results, and use currently available data and records to make qualitative analyses.

2 When electrical survey and seismic survey are used to probe water-bearing layers of interest, the interpretation of data obtained shall include the calculation of the planimetric position, dimension and buried depth of water-bearing layers of interest.

3 The interpretation of data obtained from self-potential method, mise-a-la-masse method, comprehensive logging and isotope tracer technique as used in boreholes shall include the determination of the depth, thickness, ground water tables, flow velocity, flow direction or seepage velocity of aquifers and impermeable layers.

4 In addition to being used in the analysis of formation structure and geological structure, data obtained from geothermal water resource surveys shall also be analyzed in the light of the observation that the resistivity of aquifers tends to decrease as temperature rises.

5 Data obtained from salt-to-fresh water interface survey and ground water pollution monitoring shall be used to infer resistivity of pore water, and to determine salt-to-fresh water interfaces and delineate polluted ground water areas.

6 In addition to meeting the requirements of 3.1.9 in this code, the plots of results shall also contain planimetric maps that indicate inferred water-abundant zones, water seepage zones and ground water flow direction, as well as salt-to-fresh water interfaces in areas where salt water exists, and polluted ground water areas in urban or mining regions.

4.9.4 When geophysical conditions are favorable in survey areas, the absolute error in the determination of the thickness and depth of water-bearing layers of interest and the depth of ground water table with the surface geophysics method shall be less than 20%.

4.10 Detection of Environmental Radioactivity

4.10.1 Techniques such as gamma measurement, soil radon concentration measurement and air radon concentration survey may be used for measuring the intensity of environmental gamma radiation at dam sites and main buildings, the intensity of gamma radiation from the excavation areas of foundations and underground structures, and stone pits and building materials, as well as radon concentrations in overlying soil cover in dam and main building area and concentrations of radon and its daughters in the environmental air in underground structures.

4.10.2 Investigation methods and techniques shall meet the following requirements:

  1 Gamma survey may be performed as surface gamma survey, gamma documentation and rock core gamma survey.

1) When conventional gamma radiation detectors are used for environmental detection, gamma radiation detectors shall be properly calibrated before detection is conducted.

2) For surface gamma survey, survey areas shall be selected and survey lines shall be laid out based on the results of advanced reconnaissance surveys, preferably with a line spacing of 20m to 50m and a station spacing of 10m to 20m. In abnormal sections, infilling survey lines and stations shall be used.

3) For any abnormal bedrock sections and underground caverns which are revealed by gamma surveys, gamma documentation shall be performed to delineate the abnormal area and determine its direction. For underground caverns, it is recommended that the two-wall-and-one-top or one-wall-and-one-top mode be used for gamma documentation, whereas shallow well documentation should be performed on two adjacent walls.

4) In core gamma survey, continuous gamma monitoring shall be carried out for cores, with gamma ray intensity being recorded section-by-section based on lithological properties.

2 Soil radon concentration survey shall meet the following requirements:

1) Radon concentration should be measured at each measurement station, using shallow-hole and deep-hole gas sampler respectively. During this operation, care must be taken to ensure that all stations have an equal number of gas sampling cycles and identical

measurement conditions.

2) Gas sampling-based measurements should not be used in places that are unusually moist.

3 Concentration surveys for radon and its daughters in the air shall meet the following requirements:

1) The diffusion degree of radon from rock, soil and structural zone into the air may be monitored at fixed points and times as needed, with a station spacing of 20m to 50m. In abnormal sections, infilling stations shall be used.

2) For abnormal sections, measurements shall be repeated and how well the air is circulated shall be observed and recorded.

3) For poorly ventilated underground caverns, contrast measurements shall be taken to show changes before and after ventilation is improved.

4.10.3 Data processing and interpretation shall meet the following requirements:

1 The distribution of radiation levels and sections and the relationship between it and the lithology and geological structure shall be interpreted based on the geological structure of the survey area.

2 Surface gamma survey results should be analyzed in conjunction with results from gamma measurements for outcrops, structural fractured zones and various rock samples, and from natural gamma logging.

3 The annual effective dose equivalent can be calculated from the ratio of the effective dose equivalent of gamma ray external exposure to the air absorbed dose rate of gamma radiation (which is in turn converted from air exposure rate of gamma radiation), and the annual stay time in the environment.

4 Environmental gamma radiation protection shall be based on a comprehensive approach to radiation protection which focuses on elements such as legitimation, optimization, personal dose limits (in place of the concept of threshold) and the principle of avoiding any exposure that can be avoided. GB 18871 shall be applicable as standard for radiation protection.

5 The equilibrium equivalent concentrations of radon and its daughters in the environmental air shall be calculated using relevant formulas in GB/T 14852.

6 GB 16356 shall be applicable as control standard for the equilibrium equivalent concentrations of radon and its daughters.

7 The comprehensive result plots should clearly indicate geological structure, lithology, investigation techniques, the position and intensify of anomalies and analysis results, with profiles or drill logs plotted whenever necessary.

4.10.4 To determine radioactivity measurement accuracies, relative standard deviations, by which fluctuation errors associated with statistical fluctuations can be evaluated, shall be calculated using Equation (C.5.1) given in Annex C in this code.

4.11 Quality Detection of Foundation Bedrocks

4.11.1 Techniques such as single hole acoustic wave, penetrating acoustic wave, continuous seismic wave velocity testing, shallow seismic refraction method, CT and GPR, may be used for grading foundation rock mass quality, determining the thickness of foundation rock mass unconsolidated layers and the spatial distribution of unfavorable geological bodies, determining the elevation of rock mass that may potentially be used, and evaluating and checking the quality of rock mass that has been excavated.

4.11.2 Investigation methods and techniques shall meet the following requirements:

1 Single hole acoustic wave and penetrating acoustic wave may be used for foundation rock mass quality surveys that are carried out in foundation surface survey holes.

    1) Before survey starts, the dam foundation shall be divided into a number of survey cells based on project location, geological conditions and construction schedule.

    2) Survey holes shall be arranged in groups within the survey cells, with each cell containing a suitable number of groups of drill holes, and each group having at least two holes which should be laid out at a spacing of 2m to 5m and be drilled to a depth at least 5m below the design elevation.

    3) Thickness measurements for unconsolidated layers to be blasted should be performed at the same hole site before and after blasting.

2 After excavation work has proceeded to the specified position and the foundation site has been cleared thoroughly, continuous seismic wave velocity method, shallow seismic refraction method and GPR may be used for foundation surface survey.

    1) Survey lines shall be laid out in each construction cell on the foundation surface at a line spacing of 5m to 10m and a station spacing of 1m to 3m.

    2) For foundation surfaces with no unconsolidated layers or just with thin unconsolidated layers, continuous

seismic wave velocity testing shall be conducted along survey lines and time-distance span observation system should be used. Sources may be provided by hammerings or slapping percussions in both forward and reverse direction, thereby obtaining longitudinal and traverse wave velocity data all at once.

  3) When foundation surface has thick unconsolidated layers, the shallow seismic refraction method shall be used to make refraction layering.

  4) If karst cavities are found within a suitably large space below foundation surface, the GPR shall be used.

3 If foundation rock mass has any sections that show complex local geological structure, then CT may be applied to these sections.

4 In-situ contrast tests for acoustic wave velocity and in-hole deformation modulus should be conducted in drill holes or hole sections that are representative in terms of rock mass quality. Then the relationship between acoustic wave velocity and deformation modulus should be established, taking on-site loading test records into account, and wave velocity control standard for rock mass quality should be set up.

4.11.3 Data processing and interpretation shall meet the following requirements:

1 The criterion standard of wave velocity for identifying exploitable rock mass shall be jointly defined by design, management and survey agencies based on the design requirements and the geological conditions of foundation rock mass.

2 Survey results should clearly provide information on the thickness of unconsolidated layers, the elevation of foundation surface, the quality grading of rock mass near foundation surface, and the distribution of unfavorable rock masses.

3 Contrast analysis shall be conducted for the shape, characteristics of change and integrity coefficient of single-hole and cross-hole acoustic wave velocity curves, thus obtaining the changes of the vertical and horizontal profiles of the quality of rock mass. It is also necessary to analyze the relationships between changes in the mass of rock and various elements such as formation lithology, geological structure and weathering unloading.

4 The thickness of unconsolidated layers to be blasted during construction shall be determined by changes in the acoustic wave velocity curve measured at the same location, before and after blasting, or by the velocity gradient of the acoustic wave velocity curve in the shallow hole sections.

5 When continuous seismic wave velocity testing is used, the relevant seismic wave velocity should be calculated, and the dynamic parameters of foundation rock mass can also be calculated. If single-hole acoustic wave data is available, the relationship between acoustic wave velocity and seismic wave velocity shall be provided.

6 Data obtained from the shallow seismic refraction method shall be analyzed and interpreted to provide layering information and calculate the thickness of unconsolidated layers, as well as the seismic wave velocity and integrity coefficient of unconsolidated layers and complete rock mass.

7 CT images shall be analyzed to reveal the configuration and extension range of geological structure in CT sections, and give the mass distribution of rock based on wave velocity distribution.

8 Data obtained from GPR shall provide information on the distribution and dimension of karst and unfavorable structural zones within a suitably large space below the foundation surface.

9 Statistical analysis shall be performed for velocity values by the construction cell, depth and horizontal range, from which the standard-reaching rate percentage of the wave velocity of foundation rock mass can be calculated and the elevation of foundation surface can be determined. The spatial distribution of those rock masses below foundation surface which do not reach the standard rate shall be clearly indicated.

10 In addition to meeting the requirements in the provisions of 3.1.9 in this code, the result plots should also meet the following requirements:

> 1) Statistical graphs and charts obtained from statistical analysis of wave velocity values by the cell and depth range shall be included.
> 2) Wherever in-situ static-vs-dynamic contrast testing or acoustic-vs-seismic wave velocity contrast testing is performed, relevant dependence curves shall be plotted.
> 3) Distribution map of karst and unfavorable structural zones should be included.

4.11.4 Measurement accuracies for the thickness of foundation rock mass unconsolidated layers, the spatial position of unfavorable geological bodies and the elevation of exploitable rock mass shall meet the design and construction requirements.

4.12 Detection of Grouting Effect

4.12.1 To detect the mechanical property, methods such as integrity and anti-seepage property of soil and rock masses before and after grouting, elastic wave measurement, CT, borehole deformation modulus, borehole TV, isotopic tracer, etc. may be adopted.

4.12.2 The detecting techniques and methods shall conform to the following requirements:

    1 The grouting effect detection shall start from experimental period, in which the detecting methods shall be relatively complete.

    2 The requirements for the arrangement of inspection sections or boreholes are:

        1) All holes in a grouting testing area and penetrating holes through the experimental center area shall be detected, and the number of the holes to be detected shall be no less than 5% of the total number of grouted holes.

        2) In the grouting construction stage, the numbers of inspection holes before grouting shall be larger than 1/3 of the total inspection holes.

        3) At least one group of inspection holes shall be arranged in one cell, and check cross-sections or inspection holes shall be evenly distributed with attention paid to the key parts and unusual parts.

        4) By analyzing the oversized borehole deviation, abnormal grouting process and so on, inspection holes shall be arranged at the positions that are thought to be likely to affect the grouting quality, and near the hole section that can have a large amount of grouting.

    3 To detect curtain grouting effect, such methods as tomography, borehole television, single-hole acoustic and penetrating acoustic wave shall be adopted. Isotopic tracer shall be adopted as well.

4 To detect the consolidation grouting effect, such methods as single-hole acoustic, penetrating acoustic wave and tomography shall be adopted. Borehole deformation modulus, borehole television observation and the like shall also be adopted.

5 For contrast detecting before and after grouting, such methods as penetrating acoustic wave and tomography shall be adoted, and the hole position shall be kept unchanged with the same observing system and processing technology.

6 The elasticity wave detecting should be conducted 14 days after grouting and the borehole deformation modulus 28 days after grouting:

4.12.3 Data processing and interpretation shall conform to the following requirements:

1 To analyze the detected data, as the case may be, the specific project circumstances, the contrast analysis before and after grouting, up-to-standard analysis after grouting, or phenomenon description can be chosen to evaluate the grouting effect.

> 1) The contrast analysis shall be conducted at the circumstance of comparing with the detected data in the same location before and after grouting, and calculating the increasing rate or amount.
> 2) The analysis shall be conducted at the circumstance of comparing the detected data after grouting in the grouting cell with the value of reaching rate requirements, and statistically analyzing.
> 3) Tomography may be described of the changes of low strength zone before and after grouting, and statistically analyzed in the wave velocity change of the tomography cell.

2 The opening cracks, the filling of structure fractured zones and karsts in post-grout borehole TV images shall be described.

3 The seepage velocity and flow direction of the groundwater may be calculated by the isotopic tracer method.

4 Result plots include detecting results, statistical analysis, and low strength area or seepage area distribution.

4.12.4 The accuracy rate of consolidation grouting detecting in weak parts shall be larger than 90%.

4.13 Quality Detection of Concrete

4.13.1 To detect the concrete strength, defects and the distribution of reinforcement steel in concrete, such methods as acoustic, ultrasonic-rebound synthesis, acoustic CT, borehole television, and ground penetrating radar may be adopted.

4.13.2 The detecting techniques and methods shall meet the following requirements:

1 Concrete strength should be determined by detecting concrete acoustic velocity and rebound value.

  1) With the acoustic detection method, concrete strength should be tested by establishing the correlation between acoustic velocity and compressive strength of concrete.

  2) To detect concrete surface with pairs of hanging surfaces, methods such as penetrating acoustic wave, acoustic CT, and ultrasonic rebound synthesis should be used.

  3) To detect mass concrete, the surface acoustic wave method should be adopted. When conditions permit, methods of single-hole acoustic, penetrating acoustic

wave, acoustic CT or ultrasonic rebound synthesis also can be adopted.

 4) Detecting the strength of concrete by ultrasonic rebound synthesis shall meet the requirements of JGJ/T 23.

2 Detecting concrete defects should include void, fracture extensional depth, no compact areas etc.

 1) To detect fracture extensional depth, such methods as surface acoustic wave (SAW) and penetrating acoustic wave may be used. When there are no reinforcement steel in concrete, the ground penetrating radar method may be adopted. When there are drill holes, borehole television method can be adopted.

 2) To detect concrete internal defects, such methods as acoustic CT, acoustic reflection, and ground penetrating radar should be used.

 3) Detecting concrete defects by the acoustic wave method shall conform to the provisions of CECS 21.

3 To detect the reinforcement steel distribution in concrete, ground penetrating radar method should be adopted.

4.13.3 Data processing and interpretation shall meet the following requirements:

 1 Requirements for concrete strength detecting are:

  1) To determine the relationship between concrete strength and acoustic wave velocity, a non-linear fit technique shall be used.

  2) Concrete strength of the same aggregate material but different mixture ratio may be calculated by the relationship between cement mortar velocity and concrete strength.

 2 Requirements for concrete defect detecting are:
  1) To calculate fracture depth, surface acoustic wave (SAW) shall be comparatively analyzed by such data as cross-seam and non-cross-seam detecting velocity, amplitude, and frequency. The penetrating acoustic wave method shall be based on wave velocity, first wave amplitude, frequency and other characteristics to determine fracture depth.
  2) Methods such as penetrating acoustic wave and acoustic wave CT shall be comprehensively analyzed with changes in velocity distribution, the first wave and frequency, to determine the characteristics, location and scale of internal defects.
  3) To determine the characteristics, location and size of the defect, acoustic wave reflection and ground-penetrating radar shall be based on the analysis of the frequency, phase, and continuity of reflected waves. Ground-penetrating radar shall be distinguished between concrete internal reinforcement steel distribution and abnormal defects.
 3 Requirements for detecting rebars in concrete:
  1) Location and depth of rebars shall be identified in radar images.
  2) To detect rebars and structures by radar, three-dimensional processing may be adopted.
 4 Result plots should include maps of strength distribution, defect distribution and fracture distribution, etc.
4.13.4 Detecting accuracy shall meet the following requirements:
 1 To detect the depth and range of fractures and defects,

the relative error shall be less than 10%.

2 To detect single-layer rebar position within concrete with ground penetrating radar, the absolute error shall be less than 2 cm.

4.14 Quality Detection of Concrete Lining in Cavern

4.14.1 To detect the interface contact (void condition) and cavities between concrete and its surrounding rock as well as the thickness, strength and defects of concrete lining, methods such as ground-penetrating radar, acoustic, acoustic CT and comprehensive ultrasonic rebound method may be adopted.

4.14.2 The detecting techniques and methods shall meet the following requirements:

1 Survey lines shall mainly be deployed in top arch, arching site and on both waist-line sides. The line-distance should be 1m to 5m and the station spacing should be 0.1m to 0.2m.

2 For the cavern lining without reinforcement or with a comparatively sparse single layer reinforced concrete, ground penetrating radar may be adopted to detect the lining thickness, voids, internal defects etc.

3 For the cavern lining with closer single-layer or multi-layer reinforced concrete, such methods as acoustic reflection and pulse-echo may be adopted to detect lining thickness, voids, internal defects etc. For important anomaly places, methods such as acoustic CT though holes and penetrating acoustic wave should be adopted for repeated detections.

4 To detect concrete lining strength, methods such as acoustic and ultrasonic rebound synthesis may be adopted.

Ultrasonic-rebound synthesis method to detect concrete strength shall be consistent with the relevant provisions of JGJ/T 23, and acoustic method to detect concrete strength should be consistent with the relevant provisions of 4.13 in this code.

  5 If any anomalies are found in detecting process, re-checking or infilled measuring points shall be conducted.

4.14.3 Data processing and interpretation shall meet the following requirements:

  1 Detection data shall be interpreted in combination with the design, construction and test data.

  2 Ground penetrating radar shall be used to analyze the reflected wave energy and the frequency to determine quality defects and contact status. The events of reflected wave shall be analyzed and lining thickness shall be determined by velocity parameters.

  3 For acoustic reflections, the phase, frequency, amplitude and energy attenuation of waveform, shall be analyzed to determine the lining thickness, internal defects and void status.

  4 Spectrum analysis shall be conducted in pulse-echo and lining thickness calculating. Internal defects analysis should be based on echo frequency and resonance frequency.

  5 Result plots shall include concrete strength, lining thickness, abnormal shape and range of defects.

4.14.4 Detecting accuracy shall meet the following requirements:

  1 When the conditions are favorable, detection accuracy of concrete defects (size to depth is 1 to 10) shall be more than 85%.

  2 The relative error of concrete thickness shall be less than 10%.

4.15 Detection of Relaxation Zone Around Cavern

4.15.1 To know about the stress situation of the surrounding rock in the cavity, identify the thickness of unconsolidated cavity wall, and detect the mechanical parameters of consolidated or unconsolidated rock masses, methods such as acoustic and shallow refraction wave, Rayleigh wave and seismic CT may be adopted.

4.15.2 The detection techniques and methods shall meet the following requirements:

1 Based on geophysical exploration conditions and task requirements, cross or longitudinal section detection shall be selected in one or more representative segments of the cavern. Different lithologies shall be deployed for at least one cross-section.

2 Methods such as single-hole acoustic wave and penetrating acoustic wave may be adopted for detection.

  1) The drill holes for horizontal cross-sectional detection should be deployed at the same cross-section and along the hole radial diameter axes direction, and each cross-section should be deployed for 6 – 8 holes at top of tunnel, arching site (or vertex angle) and on lumbar walls. Determining the hole depth should be based on cavity size. The hole depth shall be 3m to 15m and the standard of determining the hole depth is that the original rock stress can be reflected.

  2) Drill holes for cross or longitudinal sections may be arranged along the vital force bases as rock bolted anchor crane beams girders in the large-span and high-

side-walls of the cavern. Drill holes in top arch and lumbar walls shall be infilled.

3) The distance between drill holes for penetrating acoustic wave should be 2m to 3m, and inclination degree in a deeper drill hole shall be measured.

3 When the caverns are comparatively large, such detection methods as shallow refraction wave, Rayleigh wave may be adopted along the side walls and bottom plates of the cavities.

4 Seismic CT method may be adopted for rock walls between underground cavities.

4.15.3 Data processing and interpretation shall meet the following requirements:

1 Data of single-hole acoustic and penetrating acoustic wave method shall be based on surrounding rock in the cavern from the rock surface to the depth. Loose circles and stress distribution boundaries should be divided by the characteristics of acoustic velocity in two-dimensional profiles exhibiting a distribution regularity, which is due to the construction relaxation and stress redistribution.

2 Regions of stress decrease, stress increase and original rock stress shall be divided curve of surrounding rock wave velocities VS depth.

3 Result plots shall mainly include result profiles, as well as borehole acoustic velocity curve, Rayleigh wave dispersion curves, etc.

4.15.4 The relative error of detection for relaxed zone thickness and stress mutation location shall be less than 15%.

4.16 Quality Detection of Bolt Anchorage

4.16.1 To detect bolt length and mortar saturation, the

acoustic reflection method may be adopted.

4.16.2 The detection techniques and methods shall meet the following requirements:

1 According to the practical situation of the project, a certain amount of testing bolt should be made to experiment detection in the testing areas of same geological conditions and similar construction techniques, and the relationship between waveforms and various types of defects should be analyzed by comparison.

2 The sampling rate for detection shall be greater than 10% of the total number of bolts, and each batch shall be more than 10. When the substandard detecting number comes to more than 30% of the total number, doubled sampling rate shall be conducted. After the doubled sampling detection, if the substandard detecting number still exceeds 30% of the total, all shall be detected.

3 Sampling shall be conducted in key parts of the project (such as the underground powerhouse roof arch, rock-anchored beam), the parts of poor geological conditions (such as fracture zone, fracture-intensive zone, and loose mass of surrounding rock) and the more difficult construction of anchor bolts.

4 The shot sources should be controllable sources with wideband, short aftershocks and high repeatability. Receiver detectors should be of small-sized acceleration type. Sensitivity should be greater than 100mV/g and the band 10Hz to 2000Hz.

5 Detecting methods such as head-transmitting and receiving, side transmitting and receiving, head transmitting and side receiving can be adopted.

6 The exposed length of bolt for on-site detection should be 0.3m, whose front should be leveled and floating pulp should be cut. Detected bolt should be separated from hang nets or sprayed concrete layer, and welding points should be removed.

 7 Detection for each bolt shall be repeated three times, whose signals shall be basically the same.

4.16.3 The detection techniques and methods shall meet the following requirements:

 1 Detection record of the waveform shall be clear.

 2 To avoid confusion, the reflected signals from geological structure and bolt bottom or imperfect mortar section, end transmitting and receiving or side transmitting and receiving collected waveforms shall be analyzed by comparison.

 3 Mortar saturation should be determined by the combination of waveform characteristics and frequency characteristics.

 4 Bolt self-vibration frequency shall be determined by comparing with the same types of bolts. The lower the frequency, the more saturated grouting, the higher bond strength and the better anchoring effect.

 5 Evaluation criteria of bolt anchorage quality detection may be made according to geological conditions and design requirements of the project. If the bolt length and mortar saturation do not meet the design requirements, the bolt anchorage is considered unqualified. The qualified bolt anchorage may be further evaluated to determine its quality grade according to mortar saturation.

 6 Result plots shall be inclusive of bolt sampling location distribution maps and bolt detection result charts, of which the main result charts should include evaluation results of the original waveforms, bolt length and mortar saturation.

4.16.4 Detection accuracy shall meet the following requirements:

1 When the bolt length is less than 10m, the relative error of the detecting length shall be less than 5%. When the bolt is longer than 10m, the relative error of the detecting length shall be less than 10%.

2 When there is a single mortar defect, detecting accuracy of mortar saturation shall be greater than 90%. When there are multiple mortar defects, detecting accuracy of mortar saturation shall be greater than 80%.

4.17 Quality Detection of Cutoff Wall

4.17.1 To detect the depth, defects and uniformity of cutoff walls, such methods as resistivity imaging, CSAMT, elastic wave vertical reflection, elastic wave CT, isotope tracer, ground-penetrating radar and borehole TV observation may be adopted.

4.17.2 The detection techniques and methods shall meet the following requirements:

1 Survey lines for surface geophysical surveys shall be arranged along the wall axis. The density of stations shall be based on cutoff walls types, width and design requirements.

2 In the process of detection, if the wall depth is relatively shallow, methods as resistivity imaging and elastic wave vertical reflection should be adopted, while if the wall depth is comparatively deep, the controlled-source audio magnetotelluric (CSAMT) method should be adopted. For walls above soakage interface, the ground penetrating radar method should be adopted.

3 To further detect the wall defects, if there is any drill hole, such methods as elastic wave CT, borehole TV and isotope tracing should be adopted and the drill holes shall be deployed at the positions where anomalies are found through the surface geophysical methods.

4.17.3 The data processing and interpretation shall meet the following requirements:

1 The depth and uniformity of the cutoff walls shall be determined based on the characteristics of the physical parameter distribution in cross-section. Anomalies can be interpreted as wall defects and their locations and sizes can be determined if there is one of the following cases:

  1) Part of physical property profiles above the ground water table surface show high resistance, low acoustic velocity, and low absorption coefficient.

  2) Part of physical property profiles under the ground water table surface show low resistance, low acoustic velocity and high absorption coefficient.

  3) Part of the reflection profiles present the signal earlier than that of the wall bottom, or reflected images are not continuous.

2 Result plots should include the physical property profiles and comprehensive result interpretation map.

4.17.4 When drill holes are available, the relative error in the detection of the depth of cutoff walls shall be less than 20%.

4.18 Detection of Rockfill (Soilfill) Density and Bearing Capacity on Foundation

4.18.1 The density of rockfill mass may be determined by the additional mass method, Rayleigh wave method and nuclear densimetry method. The bearing capacity of foundation may be measured by the additional mass method as well.

4.18.2 The detection techniques and methods shall meet the following requirements:

  1 Before detecting, every type of rockfill mass in a survey area shall undergo a dig pit method additional mass method test or Rayleigh wave test, all of which shall be done at the same spot. A survey area of the same type should be tested for no less than 5 groups of tests, and additional mass test or Rayleigh wave test shall be done before the density pit test.

2 Additional mass method test and Rayleigh wave test may be adopted, when the rockfill mass is subject to layered rolling compaction, of larger grains (particle diameter$\geqslant$0.2m), and of comparatively uniform composition.

3 Nuclear density method may be adopted, when the rockfill mass is of smaller grains or composed of soil.

4 The detection instrument used in the additional mass method test shall be of wideband, high sensitivity, and moderate damping.

5 Suitable observation system should be chosen for additional mass method detection. During the detection, the additional mass m_i should be no less than 4 orders and the change of the natural frequency f_i of each order should be no less than 1Hz.

6 When Rayleigh wave detecting is adopted, the correlation coefficient between Rayleigh wave velocity and density should be set up and the geophones and seismic sources of moderate frequency should be used to ensure that the rolling quality of the inside of the top layer is effectively reflected. The shot condition set for detections should be consistent and every station should be surveyed more than 3 times to ensure that the deviation of Rayleigh wave velocity is less than 5%.

7 Nuclear density methods shall meet the provisions in SL 275.

4.18.3 Data processing and interpretation shall conform to the

following requirements:

    1 Requirements for the additional mass method are:

        1) The corresponding resonant frequency f and D shall be calculated for every order of additional mass Δm according to the time domain signals gathered and then draw the curve of $D\text{—}\Delta m$, and calculate the rigidity ratio K and Vibration mass m_0.

        2) The time-distance curves shall be plotted, and the velocity of P-wave and S-wave shall be calculated, and the wavelength λ_p shall be calculated based on $\lambda_p = v_p/f_0$ (f_0 is the resonant frequency when $\Delta m = 0$).

        3) The bearing capacity of foundation shall be calculated based on the rigidity ratio k.

        4) Direct calculated method, $k\text{—}\rho$ correlation method, and attenuation coefficient method can be adopted to calculate the station density.

    2 With the Raleigh wave method, the layer velocity of Raleigh wave shall be calculated before calculating the density according to the calibration coefficient by experiment.

    3 With the nuclear density method, the dry density of roller compacted layers shall be calculated.

    4 In result plots, the density value or the bearing capacity of foundation shall be drawn into point location maps, curves or tabulations.

4.18.4 The relative error in density detection shall be less than 5% under the condition of relatively uniform rockfill mass.

4.19 Detection of Contact Status between Steel Lining and Concrete

4.19.1 Pulse echo method and nuclear density method may be used to test voids between steel lining and concrete.

4.19.2 The detecting techniques and methods shall meet the following requirements:

 1 Survey lines shall be set at the top, middle and bottom of the steel tubes. The number of the top lines should be 3 to 5 and the station spacing should be 0.2m to 0.5m.

 2 The steel lining at stations shall be of smooth surface and be coupled well with the detectors. High frequency excitation with stable vibration energy should be used.

 3 Detection instruments shall be of broadband, high sensitivity and strong ability of frequency analysis.

 4 Each station shall be measured 3 times and the waveforms should be basically the same.

 5 Nuclear density methods shall meet the provisions in SL 275.

4.19.3 Data processing and interpretation shall meet the following requirements:

 1 With the pulse echo method, the measured waveforms and the main echo frequency shall undergo comparative analysis to determine voids and the position of voids according to the echo frequency.

 2 With the nuclear density method, the situation and range of voids shall be calculated and determined based on the integrated density, moisture content, steel plate density and concrete density.

 3 The drawing of the range of voids and the list of the positions of voids shall be made.

4.19.4 The validity of the detection of the voids between steel lining and concrete shall be more than 95%.

4.20 Quality Detection of Rockfill Dam Concrete Face Slab

4.20.1 Acoustic method, infrared thermal imaging, ultrasonic-rebound combined method and ground penetrating radar may be adopted to detect voids, internal defects and strength of concrete face slabs.

4.20.2 The detection techniques and methods shall meet the following requirements:

1 The survey lines should be of grid pattern, and line space should be 1m to 5m, station spacing 0.2m to 0.5m. During the detection, when voids or defects are found, survey lines and stations should be infilled.

2 A certain number of boreholes should be arranged to detect the anomalous points in the detection. Ultrasonic detection can be used in the testing holes to verify the strength value gained by other detecting methods.

3 When there is no or only a single layer of sparse reinforcing steel bars in the concrete face slab, the ground penetrating radar may be used to detect voids and internal defects.

4 When the concrete face slab contains a single layer of dense reinforcing steel bars or multilayer reinforcements, acoustic reflection or pulse echo methods may be adopted to detect voids and internal defects. The fracture extension in the concrete face slab surface may be detected by surface acoustic wave (SAW) or penetrating acoustic wave.

5 Acoustic methods may be also adopted to detect internal defects according to CECS 21.

6 Ultrasonic-rebound combined methods may be adopted

to detect the strength of the concrete face slab according to JGJ/T 23.

4.20.3 Data processing and interpretation shall meet the following requirements.

1 The detected abnormal data shall be interpreted both qualitatively and quantitatively based on the design of concrete face slabs and analysis of construction information.

2 If the reinforced steel bars are laid in the concrete face slabs, such abnormal situations as internal defects, steel bars and voids shall be differentiated in the radar image.

3 For acoustic method, such features of reflected signals as phase, frequency, amplitude and attenuation shall be comprehensively analyzed and recognized to determine the position of voids and defects.

4 The abnormal situation of temperature occurring in infrared imaging shall be analyzed in combination with the data gained in other methods.

5 Distribution maps of defects and strength in the concrete face slabs shall be drawn.

4.20.4 The accuracy of the position of voids and internal defects should be more than 80%.

4.21 Measurement of Hydrogeological Parameters

4.21.1 The mise-a-la-masse method, self-potential method and comprehensive logging may be used to detect the velocity, direction and seepage velocity of under groundwater, to provide the aquifer inflow and the permeability coefficient in hydrogeology test.

4.21.2 Detecting methods and techniques shall meet the

following requirements:

1 Self-potential method should be adopted to detect on the ground the direction of groundwater. In the detection, stations should be arranged in the relative flat gentle part smoothness of the survey area and ring-shaped observation should be made around the stations to measure the filtrating electrical field in different directions.

2 The mise-a-la-masse method may be adopted to detect the velocity and direction of groundwater in a water well or a single hole and the isotope tracer method may be adopted with several drill holes.

3 Well-fluid resistivity logging may be adopted to detect, in combination with natural diffusion method (salt released in the well) and isotope tracing method in boreholes, the velocity, direction and seepage velocity of groundwater. The hydraulic connection between aquifers and the aquifer inflow may be detected with borehole flowmeter or diffusion method, when the boreholes get through several aquifers of different pressures.

4 The detection of water inflow and permeability coefficient shall be carried out during pumping test or the water pressure test by measuring the axial flow rate at different depth of the borehole with the borehole flowmeter or well liquid resistivity method. The inflow of every aquifer (or the permeability rate of seepage sections) and permeability coefficient can be calculated.

5 The formal measurements should be performed when the cable is lowered in well-fluid resistivity logging. The sidewall should be flushed clean before the borehole fluid measurement is done.

4.21.3 Data processing and interpretation shall meet the following requirements:

1 The direction of groundwater in a station may be inferred from the maximum direction of the potential difference gained by the ring observation of self-potential method.

2 The flow direction of groundwater may be inferred from the maximum direction of the movement of the equipotential circle with mise-a-la-masse method. When the fixed electrodes are laid out in the direction of the upstream to survey the equipotential circle, the velocity of groundwater can be calculated according to the Equation (C. 6. 6) in Annex C. 6.

3 The aquifer inflow and permeability coefficient shall be calculated according to the axial flow rate obtained from the borehole flowmeter or well liquid resistivity method.

4 In addition to those required in the provision of 3. 1. 9 in this code, the maps or drawings shall also include the plot showing the velocity and direction of groundwater, the integrated interpretation logs and the tables for calculation results of the velocity, direction and inflow of groundwater, seepage velocity and permeability coefficient.

4.21.4 The measuring accuracy shall meet the following requirements.

1 The absolute error in the direction of groundwater shall be less than 10°.

2 The relative error in the velocity of groundwater shall be less than 20%.

3 The relative error in the inflow of groundwater shall be less than 20%.

4.22 Measurement of Rock and Soil Physical and Mechanics Parameters

4.22.1 Resistivity sounding method, acoustic method,

seismic wave method and comprehensive logging may be adopted to survey the resistivity, density, porosity, the velocity of P-wave and S-wave of soil and rock and the Poisson's ratio, dynamic elastic modulus, dynamic shear modulus, unit elastic resistance coefficient, the integrity coefficient of rock mass, anisotropy coefficient and weathering coefficient that can be calculated indirectly through the velocity of P-wave and S-wave and density.

4.22.2 The measuring methods and techniques shall meet the following requirements:

    1 Requirements for resistivity measurement are:

        1) The geotechnical resistivity may be surveyed by the resistivity sounding method and resistivity logging in a way of integrating point and area.

        2) The resistivity of bedrock may be surveyed by the resistivity sounding method in the outcrops or adits with symmetric array of small spacing.

        3) Laterolog or lateral logging may be used to survey the geotechnical resistivity in the boreholes.

    2 Density logging may be adopted to survey the density.

    3 Acoustic logging, density logging and resistivity logging may be adopted to measure porosity, which can be calculated by the following three methods:

        1) With acoustic logging, porosity can be calculated according to the Equation in Annex C.6 with the formation acoustic velocity V, the acoustic velocity of groundwater V_{pw} and the acoustic velocity of rock matrix V_{pm}.

        2) With density logging, porosity can be calculated according to the Equation in Annex C.5 with the rock

matrix density ρ_{ma}, formation bulk density ρ_b and water density in pores ρ_w.

    3) With resistivity logging, porosity can be calculated according to the equation in Annex C.6 with formation resistivity ρ_t, water resistivity in pores ρ_w and empirical coefficients m, a.

4 Different measuring methods may be adopted to measure the velocity of P-wave and S-wave in different measuring conditions.

    1) Seismic wave method may be used on the ground to measure the wave velocity of overlaying stratum and bedrock.

    2) Acoustic methods or seismic methods may be used to measure the wave velocity of rock masses in adits, shafts and underground cavities.

    3) Acoustic logging or seismic logging may be used in drill holes.

    4) Penetrating seismic wave velocity may be used to measure the velocity of P-wave and S-wave of soil and rock masses between two holes.

4.22.3 Measured parameter calculation and data interpretation shall conform to the following requirements:

1 The velocity of P-wave of a rock mass, weathering coefficient and integrity coefficient may be adopted as one of the criteria to divide rock mass weathering zones and evaluate the integrity of rock mass. The criteria shall comply to the provisions in GB/T 50287.

2 Poisson's ratio μ, dynamic elastic modulus E_d, dynamic shear modulus G_d may be calculated according to the equation in Annex C.6 with such given parameters as velocity of P-wave and

S-wave, and density. Resistance coefficient K_0 can be calculated using dynamic elastic modulus E_d and reduction factor α.

3 As to the rock mass of schistosity, bedding or lamination-like structure, the velocity of P-wave parallel to $V_p^{/\!/}$ or normal to the rock structural plane $V_p^{\perp}$ may be used to calculate the maximum anisotropy coefficient of a rock stratum according to the Equation in Annex C. 6.

4 When calculating the integrity coefficient of rock mass K_v, the standard V_{pr}, shall be correctly selected by measuring the fresh and intact rock samples (or cores) with an acoustic instrument. When there is only one type of rock mass in a survey area, a single value of V_{pr} shall be adopted by the average of a group of rock sample measurements. If there are more than two types of lithologies in one survey area, different V_{pr} shall be adopted according to the type of lithologies.

5 Weathering coefficient K_w and integrity coefficient K_V may be calculated with V_p and V_{pr} according to the equation in Annex C. 6.

4.22.4 The measuring accuracy shall meet the following requirements:

1 The absolute error between the results of density logging and indoor testing shall be less than 0.10g/cm³ in overburden and less than 0.15g/cm³ in bedrock.

2 The relative error in the porosity between the data obtained in indoor testing and those obtained with acoustic velocity logging, density logging, resistivity logging etc. shall be less than 10%.

5 Results and Reports for Geophysical Exploration

5.1 Compilation of Results and Reports

5.1.1 When a single geophysical method is used to finish one or several tasks in a work area, a single geophysical result (or special subject) report shall be compiled. When several geophysical methods are used to finish one or several tasks in a work area, a comprehensive geophysical result report shall be compiled. When the geophysical exploration for a planned phase of a project or a work area is finished, a comprehensive geophysical result report in that phase shall be compiled.

5.1.2 The single geophysical result (or special subject) report and comprehensive geophysical result report shall conform to the following requirements:

    1 The contents should include the overview, topography, the outline of geology and geophysical characteristics as well as the working methods and techniques, data interpretation and result analysis, conclusion and evaluation, and problems and suggestions.

        1) The overview includes brief introduction to the project, tasks, working hours, previous working conditions, completed workload and so on.

        2) The topography, the outline of geology and the geophysical characteristics include geographic features and geological conditions related to geophysical exploration, geological conditions (favorable and unfavorable factors) and physical characteristics.

3) Surveying methods and techniques consist of brief introduction to the principles of the methods, the arrangement of survey lines, field operating methods and techniques, instrument and equipment and working parameters.
4) Data interpretation and result analysis include evaluation of raw data, methods of data processing and interpretation, result analysis and other geological interpretation.
5) Conclusion and evaluation include the degree of the solution to the task, conclusion of the results of geophysical exploration, accuracy of the result interpretation and inspection condition of results.
6) Problems and suggestions refer to the problems still existing in this exploration and the suggestions on the exploration tasks and inspection tasks that should be supplemented; or the problems verified in the exploration and the suggestions on the practical design and construction processes.

2 The comprehensive geophysical result report shall emphasize the application of the comprehensive geophysical exploration methods to solving the geological problems and the comprehensive analysis of the obtained data from various methods.

5.1.3 Both the contents and the form of the detecting reports of the single and comprehensive geophysical exploration results are similar to the result report of geophysical exploration, but the sampling method, outline of project design and construction, evaluation of the qualification of the project, criteria of evaluation and so on, should be given.

5.1.4 Comprehensive geophysical result reports in phase shall meet the following requirements:

1 Comprehensive geophysical result reports should be compiled based on the result report of this phase or the previous phases.

2 The contents shall include overview, topography, the outline of geology, geophysical characteristics, comprehensive survey results for geophysical exploration methods, conclusion and evaluation, and problems and suggestions.

> 1) The overview includes the outline of a project, geographical location, geophysical tasks, date of start and finish of a project, arrangement of the tasks, the investigation of the comprehensive applications of various techniques and completed workload (can be listed in table).
>
> 2) Topography, the outline of geology and geophysical characteristics are geographic features related to geophysical exploration, stratigraphic structure, hydrogeology and geophysical characteristics.
>
> 3) Comprehensive survey results of geophysical exploration refer to geological-geophysical features of target formation (body), exploration contents, methods and technologies, comprehensive analysis of various methods and geological interpretation.
>
> 4) Conclusion and evaluation include the conclusion and effect of the solution to the geological problems by applying comprehensive exploration methods, evaluation of the quality and accuracy of the results.
>
> 5) Problems and suggestions consist of the problems still existing in exploration and the suggestions about the exploration tasks and verifying tasks that should be

supplemented; or the problems verified in the exploration and the suggestions about the practical design and construction processes.

5.1.5 The illustrations of the result reports may include methodological schematics, typical curves, contrast analysis plots and so on. Inserted tables should include workload table, physical parameter list, table for technical factors of instrument, tabulation of result interpretation, data sheet, accuracy table and so on.

5.1.6 Attached figures and tables in result reports shall be compliant with the requirements of each method in chapter 4 of this code.

5.2 Review of Results

5.2.1 The results shall be corrected and reviewed before submitted to the client.

5.2.2 The results that shall be corrected and reviewed are as follows:

　　1 Reports, attached figures and tables.

　　2 Such parameters as velocity, correction and interpretation data shall be interpreted and calculated.

　　3 Log book, record of instrument inspection and task statement.

　　4 The documents on technical requirements of a project from geological department, design department, supervision department or construction department.

5.2.3 The result reports shall have complete contents and diagrams and their form shall be compliant with the provisions in this code. The result data are not allowed to be reviewed in the following cases:

1 The maps or drawings that are not plotted according to the requirements.

2 The maps or drawings that are not corrected carefully and signed.

3 Data and maps are in a mess and not carefully bound.

5.2.4 If one of the following situations exists in the result reports, they will not be approved:

1 Comprehensive analysis is not complete enough, and clear conclusions and suggestions are not given for the main problems.

2 There are unclear concepts, wrong inference and incorrect conclusion.

3 The organization and structure of a report are unclear, their writing is not in correct order and there are serious mistakes in the maps and drawings.

Annex A Summary of Geophysical Exploration Applications

Table A.1 Applications for engineering geophysical exploration

Geophysical method		Investigation of overburden	Investigation of buried geological structures and fractured zones	Investigation of karst	Investigation of thickness of weathered rock mass and depth of unloading zone	Investigation of weak interlayers	Investigation of landslide	Detection of hidden defects in dams and embankments	Advanced prediction in tunnel excavation	Investigation of ground water	Thickness of underwater overburden
Electrical survey	Electrical sounding	○	△	○	○					○	△
	Electrical profiling	△	○	○	○		△	○			
	Resistivity imaging	○	○	○	○		○	○		△	
	Self-potential method		△					○		○	
	Mise-a-la-masse method										
	Induced polarization	△	△	△			○	○		○	
	CSAMT		○	○			○	○		○	
	TEM	○	○	○			○	○		○	
Ground penetrating radar	GPR	△	△	○	○			△	○	△	

Table A.1 (Continue)

Geophysical method		Investigation of overburden	Investigation of buried geological structures and fractured zones	Investigation of karst	Investigation of thickness of weathered rock mass and depth of unloading zone	Investigation of weak interlayers	Investigation of landslide	Detection of hidden defects in dams and embankments	Advanced prediction in tunnel excavation	Investigation of ground water	Thickness of underwater overburden
Seismic survey	Shallow seismic refraction	○	○	△	○		○			△	○
	Shallow seismic reflection	○	○	○	△		○		○	△	○
	Rayleigh wave method	△		△			○				
Elastic wave testing	Acoustic method	○			○	○		△			
	Seismic method	○			○			△			
Computerized tomography	Seismic wave CT	○	○	○	○						
	Acoustic wave CT	○	△	○	○						
	Electromagnetic wave CT	△	△	○	△						
Sonic echo exploration	Sonic echo exploration	○									○
Radioactivity survey	Gamma ray measurement		○							△	
	Alpha ray measurement		△							△	
	Isotopic tracer technique							○		○	

Table A.1 (Continue)

Geophysical method		Investigation of overburden	Investigation of buried geological structures and fractured zones	Investigation of karst	Investigation of thickness of weathered rock mass and depth of unloading zone	Investigation of weak interlayers	Investigation of landslide	Detection of hidden defects in dams and embankments	Advanced prediction in tunnel excavation	Investigation of ground water	Thickness of underwater overburden
Comprehensive logging	Electrical logging	○	○	○	○	○	△	△		○	
	Electromagnetic wave or radar logging	△	○	○	△		△	△		△	
	Acoustic logging	○	○	△	○	○	○				
	Radioactivity logging	△	○			○		○		○	
	Caliper		△	○	△	○	○	△			
	Borehole fluid measurement			△			○				
	Magnetic susceptibility logging		△	△			△				
	Borehole TV		○	○	○	○	△	△		○	
	Ultrasonic imaging logging		○		○	○	○			○	
	Temperature logging			△						△	

Note: ○—main method; △—Auxiliary method.

Table A.2 Applications for engineering geophysical detection

Geophysical method		Detection of Environmental radioactivity	Quality detection of foundation bedrocks	Detection of grouting effect	Quality detection of concrete	Quality detection of concrete lining in caverns	Detection of relaxation zone around caverns	Quality detection of bolt anchorage	Quality detection of cutoff wall	Detection of rockfill (soil-fill) density and bearing capacity on foundation	Detection of contact status between steel lining and concrete	Quality detection of rock-fill dam concrete face slab	Measurement of hydrogeological parameters	Measurement of rock and soil physical and mechanics parameters
Electrical survey	Electrical sounding													○
	Resistivity imaging								○					
	Self-potential method												○	
	Mise-a-la-masse method												○	
	CSAMT								○					
Ground penetrating radar	GPR		○		○	○	△		△			○		
Seismic survey	Shallow seismic refraction method		○											
	Shallow seismic reflection								○					
	Rayleigh wave method						△		△	○				○

Table A.2 (Continue)

Geophysical method		Detection of Environmental radioactivity	Quality detection of foundation bedrocks	Detection of grouting effect	Quality detection of concrete	Quality detection of concrete lining in caverns	Detection of relaxation zone around caverns	Quality detection of bolt anchorage	Quality detection of cutoff wall	Detection of rockfill (soil-fill) density and bearing capacity on foundation	Detection of contact status between steel lining and concrete	Quality detection of rock-fill dam concrete face slab	Measurement of hydrogeological parameters	Measurement of rock and soil physical and mechanics parameters
Elastic wave testing	Acoustic method		○	○	○	○	○	○	△		○	○		○
	Seismic method		○	○			△		△					○
	Seismic wave CT		△	△	△				○					
Computerized tomography	Acoustic (wave) CT		○	○	○	○			△					
	Electromagnetic wave CT		△											
Radioactivity survey	Gamma ray measurement	○												
	Alpha ray measurement	○												
	Radon concentration measurement	○												
	Isotopic tracer	△		△					○				○	

Table A.2 (Continue)

Geophysical method		Detection of Environmental radioactivity	Quality detection of foundation bedrocks	Detection of grouting effect	Quality detection of concrete	Quality detection of concrete lining in caverns	Detection of relaxation zone around caverns	Quality detection of bolt anchorage	Quality detection of cutoff wall	Detection of rockfill (soil-fill) density and bearing capacity on foundation	Detection of contact status between steel lining and concrete	Quality detection of rock-fill dam concrete face slab	Measurement of hydrogeological parameters	Measurement of rock and soil physical and mechanics parameters
Comprehensive logging	Electrical logging		○	○	○		○						○	○
	Acoustic logging		○									○		○
	Radioactivity logging (nuclear density)	△							△	○	○		△	○
	Caliper logging													
	Borehole TV			○	○	△	△		○					
	Ultrasonic imaging logging			○	○									
Additional quality method	Additional mass method									○				

Note: ○ — Main method; △ — Auxiliary method.

Annex B Table of Physical Property Parameters

Table B.1 Resistivity of common medium

Type	Description	Resistivity ρ ($\Omega \cdot m$)
Loose layer	Clay	1 to 2×10^2
	Water-bearing clay	2×10^{-1} to 1×10
	Mild clay	1×10 to 1×10^2
	Gravel plus clay	2.2×10^2 to 7×10^3
	Mild clay with gravel	8×10 to 2.4×10^2
	Pebble	3×10^2 to 6×10^3
	Water-bearing pebble	1×10^2 to 8×10^2
Sedimentary rock	Clay shale	6×10 to 1×10^3
	Sandstone	1×10 to 1×10^3
	Mudstone	1×10 to 8×10^2
	Conglomerate	1×10^2 to 1×10^4
	Limestone	6×10^2 to 6×10^3
	Marlite	1×10 to 1×10^2
	Dolomite	5×10 to 6×10^3
	Water-bearing fractured	1.7×10^2 to 6×10^3
	Anhydrite	1×10^4 to 1×10^2
	Halite	1×10^4 to 1×10^6
Metamorphic rock	Gneiss	6×10^2 to 1×10^4
	Marble	1×10^2 to 1×10^5
	Quartzite	2×10^2 to 1×10^5
	Schist	2×10^2 to 5×10^4
	Slate	1×10 to 1×10^2

Table B.1 (Contnue)

Type	Description	Resistivity ρ ($\Omega \cdot m$)
Magmatic rock	Granite	6×10^2 to 1×10^5
	Syenite	1×10^2 to 1×10^5
	Diorite	1×10^2 to 1×10^5
	Diabase	1×10^2 to 1×10^5
	Gabbro	1×10^2 to 1×10^5
	Basalt	5×10 to 1×10^5
Other	Ground water	$< 1 \times 10^2$
	River water	1×10^{-1} to 1×10^2
	Ice	1×10^4 to 1×10^8
	Karstic water	1.5×10 to 3×10
	Seawater	1×10^{-1} to 1×10

Table B.2 Density and velocity of common medium

Type	Description	Density ρ (g/cm^3)	P-wave velocity V_p (km/s)	S-wave velocity V_s (km/s)
Unconsolidated layer	Clay	1.60 to 2.04	1.2 to 2.5	0.7 to 1.4
	Damp sand	—	0.6 to 0.8	—
	Sandy clay	—	0.3 to 0.9	0.2 to 0.5
	Dry sand, gravel	—	0.2 to 0.8	0.1 to 0.5
	Water-saturated sandy gravel	—	1.5 to 2.5	—
Sedimentary rock	Conglomerate	1.90 to 2.90	1.5 to 4.2	0.9 to 2.5
	Argillaceous limestone	2.25 to 2.65	2.0 to 4.0	1.2 to 2.3
	Cherty limestone	2.80 to 2.90	4.4 to 4.8	2.6 to 3.0
	Tight limestone	2.60 to 2.77	2.5 to 6.1	1.4 to 3.5
	Shale	2.30 to 2.70	1.3 to 4.0	0.8 to 2.3

Table B. 2 (Continue)

Type	Description	Density ρ (g/cm³)	P-wave velocity V_p (km/s)	S-wave velocity V_s (km/s)
Sedimentary rock	Sandstone	2.42 to 2.77	1.5 to 5.5	0.9 to 3.2
	Dense dolomite	2.80 to 3.00	2.5 to 6.0	1.5 to 3.6
	Gypsum	2.41 to 2.58	2.1 to 4.5	1.3 to 2.8
Metamorphite	Gneiss	2.50 to 3.30	6.0 to 6.7	3.5 to 4.0
	Marble	2.68 to 2.72	5.8 to 7.3	3.5 to 4.7
	Quartzite	2.56 to 2.90	3.0 to 5.6	2.8 to 3.2
	Schist	2.68 to 3.00	5.8 to 6.4	3.5 to 3.8
	Slate	2.55 to 2.66	3.6 to 4.5	2.1 to 2.8
	Phyllite	2.71 to 2.86	2.8 to 5.2	1.8 to 3.2
Magmatite	Granite	2.30 to 2.96	4.5 to 6.5	2.4 to 3.8
	Dirorite	2.52 to 2.70	5.7 to 6.4	2.8 to 3.8
	Basalt	2.53 to 3.30	4.5 to 7.5	3.0 to 4.5
	Andesite	2.30 to 3.10	4.2 to 5.6	2.5 to 3.3
	Gabbro	2.55 to 2.98	5.3 to 6.5	3.2 to 4.0
	Diabase	2.53 to 2.97	5.2 to 5.8	3.1 to 3.5
	Olivinite	2.90 to 3.40	6.5 to 8.0	4.0 to 4.8
	Tuff	1.60 to 1.95	2.6 to 4.3	1.6 to 2.6
Others	Water	1.00	1.4 to 1.6	—
	Ice	0.80 to 0.90	3.1 to 3.6	—
	Concrete	2.40 to 2.70	2.0 to 4.5	1.2 to 2.7

Annex C Fundamental Equations and Calculation Charts

C.1 Formulae for Data Error

C.1.1 Absolute error Δ:
$$\Delta = |d_{aj} - d'_{aj}| \qquad (C.1.1)$$

C.1.2 Mean absolute error $\overline{\Delta}$:
$$\overline{\Delta} = \frac{1}{N} \sum_{i=1}^{N} |\Delta| \qquad (C.1.2)$$

C.1.3 Relative error δ:
$$\delta = \frac{|d_{aj} - d'_{aj}|}{(d_{aj} + d'_{aj})/2} \times 100\% \qquad (C.1.3)$$

C.1.4 Mean relative error $\overline{\delta}$:
$$\overline{\delta} = \frac{1}{N}\left(\sum_{i=1}^{N} \delta_i\right) \times 100\% \qquad (C.1.4)$$

C.1.5 Root mean square error m:
$$m = \sqrt{\frac{1}{N}\sum_{i=1}^{N} \delta_i^2} \times 100\% \qquad (C.1.5)$$

C.1.6 The total root mean square relative error M:
$$M = \sqrt{\frac{1}{N}\sum_{i=1}^{N} m_i^2} \times 100\% \qquad (C.1.6)$$

C.1.7 Range coefficient K:
$$K = 2\frac{d_{aj}^{\max} - d_{aj}^{\min}}{d_{aj}^{\max} + d_{aj}^{\min}} \qquad (C.1.7)$$

Where from Equation (C.1.1) to Equation (C.1.7):

N = check station, sounding station, number of lines;

d_{aj} = observation values, the mean arithmetical values of effective data from repeat observations;

d'_{aj} = observation values for system check, the mean arithmetical values of effective data from repeat observations;
d_{aj}^{max} = the maximum values in calculated data;
d_{aj}^{min} = the minimum values in calculated data.

C.2 Formulae for the Array Factor K of the Electrical Survey

C. 2. 1 Schlumberger array:
$$K = \pi \frac{AM \times AN}{MN} \quad (C.2.1)$$

C. 2. 2 Pole-dipole array:
$$K = 2\pi \frac{AM \times AN}{MN} \quad (C.2.2)$$

C. 2. 3 Bipole array:
$$K = 2\pi \times AM \quad (C.2.3)$$

C. 2. 4 Axial dipole array:
$$K = 2\pi \frac{AM \times AN \times BM \times BN}{MN(AM \times AN - BM \times BN)} \quad (C.2.4)$$

C. 2. 5 Equator dipole array:
$$K = \pi \frac{AM \times AN}{AN - AM} \quad (C.2.5)$$

C. 2. 6 Middle gradient array:
$$K = 2\pi \frac{AM \times AN \times BM \times BN}{MN(AM \times AN + BM \times BN)} \quad (C.2.6)$$

C. 2. 7 Reverse potential array for electrical logging gradient system:

1) Monopole power supply (B at the surface):
$$K = 4\pi \frac{AM \times AN}{MN} \quad (C.2.7-1)$$

2) Bipole power supply (N at the surface):
$$K = 4\pi \frac{AM \times BM}{AB} \quad (C.2.7-2)$$

Where from Equation (C.2.1) to Equation (C.2.7-2): AM, AN, MN, BM, BN, AB are all the length of electrode spacings, m.

C.3 Formulae for Electromagnetic Wave

C.3.1 Skin depth for electromagnetic sounding δ:
Under the quasi-static conditions:

$$\delta = 503\sqrt{\frac{\rho}{f}} \qquad (C.3.1)$$

Where:
ρ = resistivity, $\Omega \cdot m$;
f = frequency of electromagnetic wave, Hz.

C.3.2 The first Fresnel bandwidth of electromagnetic wave d_f:

$$d_f = \sqrt{\frac{vh}{2f}} \qquad (C.3.2)$$

Where:
v = the mean electromagnetic wave velocity, m/(ns);
h = the buried depth of an object, m;
f = frequency of electromagnetic wave, Hz.

C.3.3 The radar recording time window T:

$$T = K\frac{2H}{V} \qquad (C.3.3)$$

Where:
K = translation coefficient, 1.3 to 1.5;
H = the maximum investigation depth of radar, m;
V = the mean electromagnetic wave velocity for overlying layers, m/(ns).

C.3.4 The dipole far-field intensity of electromagnetic wave:

$$E = \frac{E_0 e^{-\beta r}}{r\cos\left(\frac{\pi}{2}\cos\theta\right)} = \frac{E_0 e^{-\beta r}}{r\cos\left(\frac{\pi}{2} \times \frac{\sqrt{r^2 - d^2}}{r}\right)} \qquad (C.3.4)$$

Where:

E_0 = the initial field intensity, V;

β = absorption coefficient, dB/m;

r = the distance between the transmitter and receiver, m;

d = horizontal distance, m;

θ = the included angle between the line and dipole, rad.

C.3.5 The wavelength of electromagnetic wave transmission in layers λ_e:

$$\lambda_e = \lambda_0 / \sqrt{\varepsilon} \qquad (C.3.5)$$

Where:

λ_0 = the wavelength of electromagnetic wave in air, m;

ε = dielectric constant of a lay.

C.4 Formulae for Seismic Survey

C.4.1 Observation system

    1 Spacing between geophone points Δx:

$$\Delta x \leqslant \frac{VT}{2\sin(i + \phi)} \qquad (C.4.1-1)$$

Where:

V = effective wave velocity, m/s;

T = apparent period of effective wave, s;

i = critical angle of refracted wave along the direction of a line;

ϕ = dip of a refractor relative to the surface (being positive in the direction of rising boundary); When the spacing between geophone points is equally spaced, Δx value in the downdip direction should be taken as standard.

    2 The number of spreads for observation system G (the value is a round number):

$$G = \frac{X_C}{(n-1)\Delta x} \qquad (C.4.1-2)$$

Where:

X_C = critical distance, i.e., length of non-traced section, m;

n = number of geophone.

C. 4. 2 Calculation of correction:

    1 The time of direct wave in a time—distance curve through correction of shot depth t_n:

$$t_n = \sqrt{\Delta t_{pn}^2 - \Delta t_s^2} \qquad (C.4.2-1)$$

Where:

Δt_{pn} = travel time of direct wave from the shot point O' to various receivers, ms;

Δt_s = the time for the corrected shot depth from the shot point O' to the surface shot point O, ms.

Other parameters see Fig. C. 4. 2.

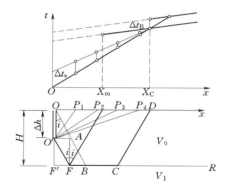

Fig. C. 4. 2 The schematic of shot depth correction

    2 The correction time of shot depth Δt_B in a refracted time-distance curve:

$$\Delta t_B = \frac{\Delta h \cos i}{V_0} \qquad (C.4.2-2)$$

Where:

Δh = depth of shot point, m;
V_0 = wave velocity of media above the shot point, m/s, see Fig. C. 4. 2.

3 The correction time Δt_D of a low velocity zone with conversion method:

$$\Delta t_D = \left(\frac{\cos i}{V_0} - \frac{\cos i}{V_1}\right)(\Delta h_Z + \Delta h_B) \qquad (C.4.2-3)$$

Where:

Δt_D = the corrected time when the low velocity zone is converted into the underlying adjacent layer, ms;
V_0 = wave velocity of low velocity zone, m/s;
V_1 = wave velocity of underlying adjacent layers in the low velocity zone, m/s.

4 The correction time Δt_D of a low velocity zone with time delay method:

$$\Delta t_D = \frac{\cos i}{V_0}(\Delta h_Z + \Delta h_B) \qquad (C.4.2-4)$$

Where:

Δh_Z = thickness of a low velocity zone where a geophone is located, m;
Δh_B = thickness of a low velocity zone at a shot point, m;
i = the critical angle of between the (low velocity zone and refracted wave in the underlying adjacent layer, rad.

5 Topographic correction Δt_C:

$$\Delta t_C = \frac{\cos i}{V_0}(\Delta h_Z + \Delta h_B) \qquad (C.4.2-5)$$

Where:

Δh_Z = vertical distance from geophone points to correction lines, m;
Δh_B = vertical distance from shot points to correction lines, m;
V_0 = wave velocity of near-surface media, m/s;

i = critical angle between near-surface media and refracted waves in underlying adjacent layers, rad.

C.4.3 Calculation of wave velocity:

1 The mean wave velocity:

$$\overline{V} = \frac{h_1 + h_2 + \cdots + h_n}{\dfrac{h_1}{V_1} + \dfrac{h_2}{V_2} + \cdots + \dfrac{h_n}{V_n}} = \frac{\sum\limits_{i=1}^{n} h_i}{\sum\limits_{i=1}^{n} t_i} \quad (C.4.3-1)$$

Where:
h_i = thickness of various layers, m;
t_i = travel time of elastic waves vertically penetrating various layers, ms.

2 Effective velocity is derived from the cross point method in refraction time-distance curves:

$$V_{en} = \frac{X_{cn}}{t_{cn}} \quad (C.4.3-2)$$

Where:
V_{en} = effective velocity of layers above the refraction interface of the nth layer, m/s;
X_{cn} = critical distance of the refracted waves of the nth layer, m;
t_{cn} = travel time to the critical point of the nth layer, ms.

3 Effective velocity are derived from the square coordinated method from reflection time-distance curves:

1) Effective velocity is derived from single shooting reflection time-distance curves with the $t^2 - x^2$ coordinated method, applicable to the interface dip of less than 15°.

$$V_{ef} = \sqrt{\frac{\Delta X}{\Delta T}} \quad (C.4.3-3)$$

Where:
$X = x^2$, $T = t^2$ (see Fig. C.4.3-1).

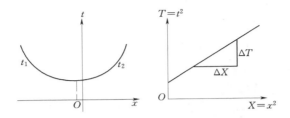

Figure C. 4. 3 - 1 Effective velocity of reflected waves is derived from the square coordinated method

2) Effective velocity is derived from a forward and reverse shooting reflection time-distance curves with the $t^2 - x^2$ coordinated method, applicable to the tilted interface.

$$V_{ef} = \sqrt{\frac{\Delta X}{\Delta T}} \qquad (C.4.3-4)$$

$$X = x^2, \quad T = t^2 = \frac{t_1^2 + t_2^2}{2}$$

Where:

t_1, t_2 = separately reflected wave times of geophone points with the equal shot-geophone distance on both side of a shot point, ms.

3) Effective velocity is derived from reflection wave forward and reverse T—D curves with the $u-x$ coordinated method, applicable to the interface dip of less than 7° (see Fig. C. 4. 3 - 2 for parameters).

$$V_{ef} = \sqrt{2L \frac{\Delta x}{\Delta u}} \qquad (C.4.3-5)$$

$$u = t_1^2 - t_2^2$$

Where:

L = distance of reflected waves from a seismic source to the farthest geophone point, m.

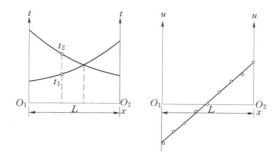

**Figure C. 4. 3 – 2   Effective velocity of reflected waves
is derived from the difference
time-distance curves**

C. 4. 4   The calculation of seismic interfaces:

1   Thickness of various refractors is calculated from a forward or reverse shooting time-distance curves with the intercept time:

$$\left.\begin{array}{l} h_1 = \dfrac{t_{01}}{2} \dfrac{V_1 V_2}{\sqrt{V_2^2 - V_1^2}} \\ \cdots \\ h_n = \dfrac{t_{0n}}{2} \dfrac{V_n V_{n+1}}{\sqrt{V_{n+1}^2 - V_n^2}} - \sum_{k=1}^{n-1} h_k \dfrac{V_n \sqrt{V_{n+1}^2 - V_k^2}}{V_k \sqrt{V_{n+1}^2 - V_n^2}} \end{array}\right\}$$

(C. 4. 4 – 1)

Where:

V_1, V_2, V_3, $\cdots$, V_n = wave velocity of various media, m/s;

t_{01}, t_{02}, t_{03}, $\cdots$, t_{0n} = intercept time of refracted waves in various layers, ms (see Fig. C. 4. 4 – 1).

2   Thickness of various refractors is calculated from a forward or reverse shooting time-distance curves with the critical distance:

$$h_1 = \dfrac{X_{12}}{2} \sqrt{\dfrac{V_2 - V_1}{V_2 + V_1}}$$

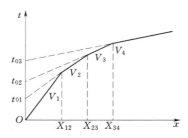

Figure C. 4. 4 – 1 Refraction interface is derived from the intercept time and critical distance

...

$$h_n = \frac{X_{n(n+1)}}{2}\sqrt{\frac{V_{n+1}-V_n}{V_{n+1}+V_n}} - \sum_{k=1}^{n-1} h_k \frac{V_n\sqrt{V_{n+1}^2-V_k^2} - V_{n+1}\sqrt{V_n^2-V_k^2}}{V_k\sqrt{V_{n+1}^2-V_n^2}}$$

(C. 4. 4 – 2)

Where:

V_1, V_2, V_3, ···, V_{n+1} = wave velocity of various media, m/s;

X_{12}, X_{23}, X_{34}, ···, $X_{n(n+1)}$ = critical distance of refracted waves in various layers, m. See Fig. C. 4. 4 – 1.

3 Depth of refraction interfaces is derived from a forward and reverse shooting time-distance curves with t_0 method:

$$t_0(x) = t_A(x) - [T_{AB} - t_B(x)] \quad \text{(C. 4. 4 – 3)}$$

$$\theta(x) = t_A(x) + [T_{AB} - t_B(x)] \quad \text{(C. 4. 4 – 4)}$$

$$V_2 = \frac{2\Delta x}{\Delta\theta(x)}\cos\phi \approx \frac{2\Delta x}{\Delta\theta(x)} \text{(when } \phi \text{ is less than} 15°)$$

(C. 4. 4 – 5)

$$h(x) = \frac{V_1 t_0(x)}{2\cos i} \quad \text{(C. 4. 4 – 6)}$$

Where:

$t_A(x)$ = observation time of forward time-distance curves at the shot point A, ms;

$t_B(x)$ = observation time of forward time-distance curves at the shot point B, ms;

T_{AB} = reciprocal time, ms;

V_1 = the mean wave velocity or effective velocity of the overlying layer above the refraction interface, m/s;

V_2 = slide wave velocity of refraction interfaces, m/s;

$h(x)$ = depth of refraction interfaces, m.

4 Depth of refraction interface is derived from a forward and reverse shooting T—D curves with the time-delay method:

$$D(x) = \frac{t_A(x) + t_B(x)}{2} - \frac{T_{AB}}{2} \qquad (C.4.4-7)$$

$$T'(x) = t_A(x) - D(x) \qquad (C.4.4-8)$$

$$V_2 = \frac{\Delta x}{\Delta T'(x)} \qquad (C.4.4-9)$$

$$h(x) = \frac{V_1 D(x)}{\cos i} \qquad (C.4.4-10)$$

5 Refraction interface is derived from a forward and reverse shooting T—D curves with the conjugate point method. See Fig. C.4.4-2. The interpretation procedure for the conjugate point method is as follows:

1) When the mirror time-distance curve t'_{BR} is drawn from the forward and reverse shooting T—D curve t_{BR}, allowing for the time-distance curve $t'_{BR} = T_{AB} - t_{BR}$.

2) The cross point method is used to derive effective velocity V_1, drawing the $(t_{AR} + t'_{BR})/2$ curve, i.e., the interface velocity V_2 as the inverse slope of a point-linked line in the time-distance curve t_{AR} and the mirror time-distance curve t'_{BR}.

3) Take a point b in the t'_{BR} curve, with $\dfrac{1}{V_1 \sin i}$ as the slope to draw a ray that is crossed with the t_{AR} curve at

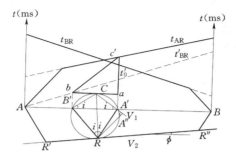

Figure C. 4. 4 – 2 The schematic for use of the conjugate point method for obtaining the refraction interfaces

the c' point, then, the x-coordinate of b and c' points is a pair of conjugate points B' and A', assuming $A'B' = d$ and $t_{AR}(c') - t'_{BR}(b) = t_0$.

4) Draw a ray that is crossed with x-coordinate at the C point forming the I angle and separately passing through A' and B'.

5) With C as the center of a circle and $CR = \dfrac{V_1 t'_0}{2\cos i}$ as the radius, draw an arc tangent to a refraction interface.

6) Repeat the above course every appropriate distance on the mirror curve t'_{BR}, which can result in the continuous plot of an interface.

6 Refraction interface is derived from the forward and reverse T—D curve with the time-field method.

The sum of time for any points in a refraction interface R corresponding to the equal travel time surface of two encountered refractions (events) is equal to the reciprocal time T. With the actually observed time values and the mean or effective velocity V_1 in overlying layers as the time field of two encountered refractions, the point-linked line that meet the relationship $t_A + t_B = T$, is the traced interface. The interface velocity is:

$$V_2 = \frac{\Delta \zeta}{\Delta t} \qquad (C.4.4-11)$$

Where:

$\Delta \zeta =$ the interface distance between two equal travel time surfaces, m;

$\Delta t =$ the time difference between two equal travel time surfaces, ms.

7 Depth of refraction interfaces is derived from composite time-distance curve with the layering method:

$$\left. \begin{aligned} h_1 &= \frac{V_1 t_{01}}{2\cos i_{12}} \\ &\cdots \\ h_n &= \frac{V_n t_{0n}}{2\cos i_{n(n+1)}} - \sum_{k=1}^{n-1} h_k \frac{\cos i_{k(n+1)}}{\sin i_{kn} \cdot \cos i_{n(n+1)}} \end{aligned} \right\} \qquad (C.4.4-12)$$

$$i_{kn} = \sin^{-1} \frac{V_k}{V_n}$$

Where:

$V_1, V_2, V_3 \cdots V_n =$ wave velocities in various media, m/s;

$t_{01}, t_{02}, t_{03} \cdots t_{0n} = t_0$ time of refracted wave in various layers, ms.

8 Depth of reflection interfaces is derived from expanding spreads:

$$H = \frac{\sqrt{(vt)^2 - X^2}}{2} \qquad (C.4.4-13)$$

Where:

$v =$ the mean wave velocity or effective velocity, m/s;

$t =$ travel time of reflected waves, ms;

$X =$ shot-geophone distance, m.

9 Depth of reflection interfaces is derived from the common-offset profile:

1) The circle method:

$$H = \frac{\sqrt{(vt)^2 - L^2}}{2} \quad \text{(C.4.4-14)}$$

Where:

L = offset distance, m.

2) The ellipse method:

The parameter equation for reflection interfaces with the second half elliptical trace determined by the elliptic equation is:

$$\frac{X'^2}{a^2} + \frac{Y'^2}{b^2} = 1 \quad \text{(C.4.4-15)}$$

Where:
$$X' = X - \frac{L}{2} = X - C$$

$$Y' = Y = b\sqrt{1 - \left(\frac{X-C}{a}\right)^2}$$

$$a = \frac{1}{2}vt$$

$$b = h = \frac{\sqrt{(vt)^2 - L^2}}{2}$$

$$c = \frac{L}{2}$$

Assuming that the step-length from O to L at X is the geophone interval ΔX, calculate the second-half elliptical trace and use the envelopes to construct a reflection interface (see Fig. C.4.4-3 for parameters).

10 Depth of reflection interfaces is derived from the common-depth-point stack time section (profile).

$$H = \frac{vt}{2} \quad \text{(C.4.4-16)}$$

Where:

v = the mean velocity or stacking velocity, m/s;

t = the corrected travel time of reflected waves, ms.

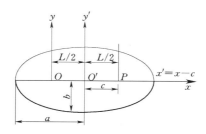

Figure C. 4. 4 – 3 The schematic diagram for use of the ellipse method of obtaining a reflection interface

Drawing an arc with the common midpoint as the center of circle and H as the radius, the envelopes are constructed as a reflection interface.

C. 4. 5 Calculation of Rayleigh wave:

1 The dominant frequency f_0 for weight-drop:

$$f_0 = \frac{1}{2\pi}\sqrt{\frac{4G_d r_0}{M(1-\mu)}} \qquad \text{(C. 4. 5 - 1)}$$

Where:

G_d = shear modulus, Pa;

r_0 = basal area, m²;

M = weight, N;

μ = Poisson ratio.

2 Estimates of the frequency point step-length Δf with the steady-state method.

$$\Delta f = (2f^2/V_R)\Delta H \qquad \text{(C. 4. 5 - 2)}$$

Where:

f = testing frequency, Hz;

V_R = Rayleigh wave velocity, m/s;

ΔH = incremental depth, m.

3 Calculation of Rayleign wave velocity V_R with the time-difference method.

$$V_R = \Delta x / \Delta t \qquad (C.4.5-3)$$

Where:

Δx = geophone (detection) interval, m;

Δt = time difference, s.

4 Calculation of Rayleigh wave velocity V_R with phase difference.

$$V_R = 2\pi f \Delta x / \Delta \phi \qquad (C.4.5-4)$$

Where:

Δx = spacing, m;

$\Delta \phi$ = phase difference, rad.

5 Calculation of relevancy $r(k)$ with the cross correlation method.

$$r(k) = \frac{1}{N} \sum_{l=0}^{N-1} e(l) g(l+k) \quad (k = 0, 1, \cdots, N-1)$$

$$(C.4.5-5)$$

Where:

$e(l)$, $g(l)$ = correlation function with some similarity.

6 The empirical equation for S-wave V_S and Rayleigh wave velocity V_R:

$$V_R = \frac{0.87 + 1.12\mu}{1 + \mu} \times V_S \qquad (C.4.5-6)$$

Where:

V_S = S-wave velocity, m/s.

7 Layer velocity of Rayleigh wave:

    1) When the mean formation velocity is increasing with the depth, V_{Rn} for the $n-1$ to n layer is:

$$V_{Rn} = \frac{H_n \overline{V}_{Rn} - H_{n-1} \overline{V}_{Rn-1}}{H_n - H_{n-1}} \qquad (C.4.5-7)$$

    2) When the mean formation velocity is decreasing with

the increasing depth, V_{Rn} for the $n-1$ to n layer is:

$$V_{Rn} = \frac{H_n - H_{n-1}}{H_n/\overline{V}_{Rn} - H_{n-1}/\overline{V}_{Rn-1}} \quad (C.4.5-8)$$

3) When not considering the variation trend of the mean formation velocity with depth, V_{Rn} for the $n-1$ to n layers is:

$$V_{Rn} = \frac{\overline{V}_{Rn}^2 H_n - \overline{V}_{Rn-1}^2 H_{n-1}}{H_n - H_{n-1}} \quad (C.4.5-9)$$

Where from Equation (C.4.5-7) to Equation (C.4.5-9):

H_n = depth at the nth point, m;

H_{n-1} = depth at the $n-1$ point, m;

$\overline{V}_{Rn}$ = the mean Rayleign wave velocity above the depth at the nth point, m/s;

$\overline{V}_{Rn-1}$ = the mean Rayleign wave velocity above the depth at the $n-1$ point, m/s;

V_{Rn} = the interlayer Rayleigh wave velocity in the depth H_n to H_{n-1}, m/s.

C.5 The Formulae for Radioactivity Survey

C.5.1 Error of radioactivity statistical fluctuations

1 The standard error for pulse counters, σ:

$$\sigma = \pm\sqrt{N} \quad (C.5.1-1)$$

2 The relative standard error for pulse counters, δ:

$$\delta = \pm\frac{\sigma}{N} = \pm\frac{1}{\sqrt{N}} = \pm\frac{1}{\sqrt{Nt}} \quad (C.5.1-2)$$

3 The standard error for count-rate type radiometers, σ:

$$\sigma = \pm\sqrt{\frac{n}{2RC}} \quad (C.5.1-3)$$

4 The relative standard error for count type radiometers, δ:

$$\delta = \pm \frac{1}{\sqrt{2\bar{n}RC}} \quad (C.5.1-4)$$

Where from Equation (C. 5. 1 - 1) to Equation (C. 5. 1 - 4):

$\bar{N}$ = the mean reading;

N = number of pulses;

t = reading time, s;

$RC = \tau$ = time constant of integral circuit;

$\bar{n}$ = count rate of pulses, i. e., pulse number in unit time.

C. 5. 2 Calculation of permeability coefficient K with isotopic tracer logging.

$$K = \frac{V_f}{I}$$

$$V_f = \frac{\pi(\gamma^2 - \gamma_0^2)}{2\alpha rt} \ln \frac{N_0 - N_b}{N_t - N_b} \quad (C.5.2)$$

Where:

I = hydraulic gradient of underground water near the test borehole;

γ = internal radius of filter pipe in the test borehole, m;

γ_0 = radius of probe, m;

t = time needed for changes in the tracer density from N_0 to N_t, d;

N_0 = the initial count rate of isotope in a hole;

N_t = count rate of isotope at the time of t;

N_b = background count rate of radioactivity;

α = correction coefficient for flow field distortion.

C. 5. 3 Porosity is derived from density log:

$$n = \frac{\rho_{ma} - \rho_b}{\rho_{ma} - \rho_w} \quad (C.5.3)$$

Where:

ρ_{ma} = matrix density of rocks, g/cm³;

ρ_b = formation bulk density, g/cm³;
ρ_w = water density in pores, g/cm³.

C.6 The Formulae for Rock Mass Mechanics Parameters

C.6.1 Density is derived from the additional mass method:

$$\left. \begin{array}{l} \rho = \dfrac{m_0}{A\dfrac{V_p}{2\beta f_0}}, \quad m_0 = \dfrac{1}{K'} = KD_0 \\ k = \dfrac{\Delta m}{\Delta D}, \quad D = \dfrac{1}{(2\pi f)^2} \end{array} \right\} \quad (C.6.1)$$

Where:

ρ = rockfill density, g/cm³;

A = area of bearing plate, m²;

V_p = wave velocity of measuring points, m/s;

β = attenuation coefficient (through comparison experiment between the pit-dig method and additional mass);

m_0 = mass of vibration, kg;

f_0 = intercept natural vibration frequency when Δm is 0, Hz;

k = rigidity;

Δm = graded weight of additional mass, kg. m_0 is solved as per analysis. m_0 is equal to the absolute value of intercept of ($D-\Delta m$) curve at Δm coordinate axis (see Fig. C.6.1).

C.6.2 Calculation of depth H of reflection interfaces with pulse echo:

$$H = \dfrac{V_p}{2f_0} \quad (C.6.2)$$

Where:

V_p = acoustic P-wave velocity, m/s;

f_0 = echo frequency, Hz.

C.6.3 Relevant mechanics parameters:

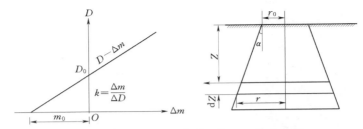

Figure C.6.1 Mass of vibration m_0 is derived from the graphical method

1 Poisson ratio μ:

$$\mu = \frac{V_p^2 - 2V_s^2}{2(V_p^2 - V_s^2)} \quad \text{(C.6.3-1)}$$

2 Dynamic elasticity modulus E_d:

$$E_d = V_p^2 \rho \frac{(1+\mu)(1-2\mu)}{(1-\mu)} \quad \text{(C.6.3-2)}$$

$$E_d = 2V_s^2 \rho (1+\mu) \quad \text{(C.6.3-3)}$$

3 Dynamic shear modulus G_d:

$$G_d = V_s^2 \rho \quad \text{(C.6.3-4)}$$

4 Unit elastic resistance coefficient K_0:

$$K_0 = \frac{E_d}{100(1+\mu)} \alpha \quad \text{(C.6.3-5)}$$

5 Anisotropy coefficient η:

$$\eta = \frac{V_p^{/\!/}}{V_p^{\perp}} \quad \text{(C.6.3-6)}$$

6 Weathering coefficient K_w:

$$K_w = V_p / V_{pr} \quad \text{(C.6.3-7)}$$

7 Integrity coefficient K_v:

$$K_v = (V_p / V_{pr})^2 \quad \text{(C.6.3-8)}$$

Where:

V_p = P-wave velocity of a rock mass, m/s;

V_s = S-wave velocity of a rock mass, m/s;

ρ = density of a rock mass, g/cm³;

α = reduction factor;

$V_p^{//}$ = P-wave velocity parallel to the direction of a rock mass structure plane, m/s;

$V_p^{\perp}$ = P-wave velocity normal to the direction of a rock mass structure plane, m/s;

V_{pr} = compression wave velocity of fresh complete rock mass, m/s.

C. 6. 4 Porosity n is derived from acoustic log:

$$n = \frac{1/V - 1/V_{pm}}{1/V_{pw} - 1/V_{pm}} \quad (C.6.4-1)$$

$$n = \frac{t - t_{pm}}{t_w - t_{pm}} \quad (C.6.4-2)$$

Where:

V_{pw} = acoustic wave velocity of water in a rock mass, m/s;

V_{pm} = acoustic wave velocity of rock matrix, m/s;

V = formation acoustic velocity, m/s;

t = time for the acoustic wave transmission for 1m in a rock mass, ms;

t_{pm} = time for the acoustic wave transmission for 1m in rock matrix, ms;

t_w = time for the acoustic bulk transmission for 1m in water, ms.

C. 6. 5 Porosity is derived from resistivity log:

$$n = \sqrt[m]{a\rho_w / \rho_t} \quad (C.6.5)$$

Where:

ρ_t = formation resistivity, $\Omega \cdot m$;

ρ_w = water resistivity in pores, $\Omega \cdot m$;

m, α = empirical factor (For the unconsolidated layer with high porosity, $\alpha = 0.62/m = 2.15$ or $\alpha = 0.81/m = 2.00$. For limestone, $\alpha = 1.00/m = 2.00$ to 2.50. For fracture-developed limestone, $\alpha=1.00/m=1.12$ to 1.30).

C. 6. 6 Calculation of groundwater flow velocity using the displacement of equipotential circle and time with the Mise-a-la-masse method:

$$v = \Delta R_i / \Delta t_i \quad (C.6.6-1)$$
$$v_j = v/\cos\beta \quad (C.6.6-2)$$

Where:

$v=$ groundwater flow velocity, m/h;

$v_j=$ groundwater flow velocity through topographic correction, m/h;

$\Delta R_i=$ displacement of upward equipotential circle of groundwater flow, m;

$\Delta t_i=$ time interval between two observations of equipotential circle, h;

$\beta=$ topographic slope.